Preface

Primary Mathematics (U.S. Edition) is a complete program based on the high... **Mathematics** series from Singapore. Designed to equip students with a strong foundation in ... topics are covered in depth and taught to mastery. By focusing on mathematical understanding, the program aims to help students develop logical thinking and problem solving skills.

The Core Curriculum for each semester comprises one Textbook and one Workbook. The textbook is a non-consumable class activity book and is used during class instruction. The workbook is consumable and is used for independent work.

The **Primary Mathematics (U.S. Edition)** program calls for direct instruction and focuses on mathematical thinking with immediate application of new skills to problem solving. By encouraging students to solve problems in a variety of ways, this program promotes an understanding of the way mathematical processes work.

The **Primary Mathematics (U.S. Edition)** program follows a Concrete → Pictorial → Abstract approach. This enables students to encounter math in a meaningful way and translate mathematical skills from the concrete to the abstract. It allows students to understand mathematical concepts before learning the rules of formulaic expressions. Students first encounter the mathematical concepts through the use of manipulatives. They then move on to the pictorial stage in which pictures are used to model problems. When they are familiar with the ideas taught, they progress to a more advanced or abstract stage in which only numbers, notation, and symbols are used.

The **Primary Mathematics Teacher's Guide** is designed for teachers new to the program. It provides background notes on the mathematical concepts for each unit and relates the concepts being learned in a specific unit to the program as a whole. It includes suggestions for presenting the concepts concretely and for differentiating instruction. It also provides detailed lesson plans and some additional suggested problems. Answers are provided for all problems in both the textbook and workbook, and often include suggested solutions (not all possible methods of solution are given). The appendix includes Mental Math worksheets and other aids. It is hoped that this guide will increase the teacher's understanding of mathematics in general, as well as how it is taught in this program.

The **Primary Mathematics Teacher's Guide** is a *guide*. Although the suggested lessons are detailed, each classroom is unique, and each teacher is unique. One guide cannot anticipate all possible situations. Experienced teachers can and should bring their own methods and experiences and strengths to the classroom in teaching the concepts to the students.

It is important that students understand concepts and not just follow procedures, but they still must have adequate practice in following the standard procedures correctly, such as the addition and subtraction algorithms. Following a procedure alone is not sufficient if the underlying concepts are not understood. By requiring students to apply their knowledge in new situations, you can determine whether they understand the concepts. It is also important that students learn to reason through word problems and solve them logically, rather than being required to follow a step-by-step procedure for problem solving that can only apply to problems that lend themselves to those specific steps. Otherwise, students will not learn to reason through problems that do not work well with a given predetermined set of steps. Strategies that can apply to many types of problems are more valuable than strategies that apply to only a few types of problems, or only to easier problems.

Allow students sufficient use of manipulatives so that they are seen as tools, rather than toys, and can use them appropriately.

Table of Contents

Developmental Continuum

Primary Mathematics 1A
1 Numbers 0 to 10
  1 Counting
2 Number Bonds
  1 Making Number Stories
3 Addition
  1 Making Addition Stories
  2 Addition With Number Bonds
  3 Other Methods of Addition
4 Subtraction
  1 Making Subtraction Stories
  2 Methods of Subtraction
5 Ordinal Numbers
  1 Naming Position
6 Numbers to 20
  1 Counting and Comparing
  2 Addition and Subtraction
7 Shapes
  1 Common Shapes
8 Length
  1 Comparing Length
  2 Measuring Length
9 Weight
  1 Comparing Weight
  2 Measuring Weight

Primary Mathematics 2A
1 Numbers to 1000
  1 Looking Back
  2 Comparing Numbers
  3 Hundreds, Tens and Ones
2 Addition and Subtraction
  1 Meanings of Addition and Subtraction
  2 Addition Without Renaming
  3 Subtraction Without Renaming
  4 Addition With Renaming
  5 Subtraction With Renaming
3 Length
  1 Measuring Length in Meters
  2 Measuring Length in Centimeters
  3 Measuring Length in Yards and Feet
  4 Measuring Length in Inches
4 Weight
  1 Measuring Weight in Kilograms
  2 Measuring Weight in Grams
  3 Measuring Weight in Pounds
  4 Measuring Weight in Ounces
5 Multiplication and Division
  1 Multiplication
  2 Division
6 Multiplication Tables of 2 and 3
  1 Multiplication Table of 2
  2 Multiplication Table of 3
  3 Dividing by 2
  4 Dividing by 3

Primary Mathematics 3A
1 Numbers to 10,000
  1 Thousands, Hundreds, Tens and Ones
  2 Number Patterns
2 Addition and Subtraction
  1 Sum and Difference
  2 Adding Ones, Tens, Hundreds and Thousands
  3 Subtracting Ones, Tens, Hundreds and Thousands
  4 Two-Step Word Problems
3 Multiplication and Division
  1 Looking Back
  2 More Word Problems
  3 Multiplying Ones, Tens and Hundreds
  4 Quotient and Remainder
  5 Dividing Hundreds, Tens and Ones
4 Multiplication Tables of 6, 7, 8 and 9
  1 Looking Back
  2 Multiplying and Dividing by 6
  3 Multiplying and Dividing by 7
  4 Multiplying and Dividing by 8
  5 Multiplying and Dividing by 9
5 Money
  1 Dollars and Cents
  2 Addition
  3 Subtraction

Primary Mathematics 1B
1 Comparing Numbers
  1 Comparing Numbers
  2 Comparison by Subtraction
2 Graphs
  1 Picture Graphs
3 Numbers to 40
  1 Counting
  2 Tens and Ones
  3 Addition and Subtraction
  4 Adding Three Numbers
4 Multiplication
  1 Adding Equal Groups
  2 Making Multiplication Stories
  3 Multiplication Within 40
5 Division
  1 Sharing and Grouping
6 Halves and Quarters
  1 Making Halves and Quarters
7 Time
  1 Telling Time
8 Numbers to 100
  1 Tens and Ones
  2 Order of Numbers
  3 Addition Within 100
  4 Subtraction Within 100
9 Money
  1 Bills and Coins
  2 Shopping

Primary Mathematics 2B
1 Addition and Subtraction
  1 Finding the Missing Number
  2 Methods for Mental Addition
  3 Methods for Mental Subtraction
2 Multiplication and Division
  1 Multiplying and Dividing by 4
  2 Multiplying and Dividing by 5
  3 Multiplying and Dividing by 10
3 Money
  1 Dollars and Cents
  2 Adding Money
  3 Subtracting Money
4 Fractions
  1 Halves and Quarters
  2 Writing Fractions
5 Time
  1 Telling Time
  2 Time Intervals
6 Capacity
  1 Comparing Capacity
  2 Liters
  3 Gallons, Quarts, Pints and Cups
7 Graphs
  1 Picture Graphs
8 Geometry
  1 Flat and Curved Faces
  2 Making Shapes
9 Area
  1 Square Units

Primary Mathematics 3B
1 Mental Calculation
  1 Addition
  2 Subtraction
  3 Multiplication
  4 Division
2 Length
  1 Meters and Centimeters
  2 Kilometers
  3 Yards, Feet and Inches
  4 Miles
3 Weight
  1 Kilograms and Grams
  2 More Word Problems
  3 Pounds and Ounces
4 Capacity
  1 Liters and Milliliters
  2 Gallons, Quarts, Pints and Cups
5 Graphs
  1 Bar Graphs
6 Fractions
  1 Fraction of a Whole
  2 Equivalent Fractions
7 Time
  1 Hours and Minutes
  2 Other Units of Time
8 Geometry
  1 Angles
  2 Right Angles
9 Area and Perimeter
  1 Area
  2 Perimeter
  3 Area of a Rectangle

Material

It is important to introduce the concepts concretely, but it is not important exactly what manipulative is used. Teachers need to use equivalent material that can be displayed on the board, either by using an overhead projector or objects with magnetic or sticky backs.

Whiteboard and dry-erase Markers
An individual whiteboard for each student allows them to work out the problems presented during a lesson. Each student can then hold up the board when he or she has finished the problem.

Place-value discs
Round discs with 0.001, 0.01, 0.1, 1, 10, or 100 written on them. You can label round counters using a permanent marker. You need 20 of each kind for each student.

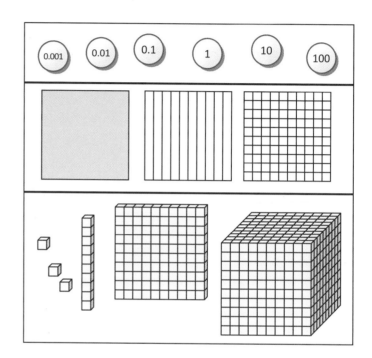

Fraction squares (paper)
A whole square and squares divided into tenths and hundredths. Copy the ones in the appendix. Base-10 block flats, rods, and unit cubes can be used instead.

Base-10 blocks
A set usually has 100 unit-cubes, 10 or more ten-rods, 10 hundred-flats, and 1 thousand-block.

Number cubes
Cubes that you can label and throw, like dice. These are for group games. You need four per group.

Multilink cubes
Cubes that can be linked together on all 6 sides to make 3-dimensional solids.

Centimeter cubes
Cubes 1 centimeter on a side.

Graph papers
Centimeter graph paper, square dot paper, isometric dot paper. These can be copied from the appendix of this guide.

Measuring tools
Meter sticks and yard sticks
Rulers (12 inch/30 cm) for each student
Beakers, graduated cylinders, scale (optional)
Set squares (plastic triangles with one right angle)
Liter measuring cup
Dropper or teaspoon

Supplementary Math

Singaporemath.com carries a number of supplementary workbooks that can be used for extra practice or enrichment, including *Extra Practice for Primary Mathematics* (U.S. Edition) and *Primary Mathematics Intensive Practice* (U.S. Edition). These can be used as a source of more problems or more challenge.

1 Decimals

Objectives

- Read and write decimal numbers of up to 3 places.
- Locate decimals on a number line.
- Compare and order decimals and simple fractions.
- Relate money in dollars and cents to decimals.
- Convert a decimal number to a fraction.
- Convert a fraction with a denominator that is a factor of 100 to a decimal.
- Add and subtract tenths and hundredths.
- Round decimal numbers to the nearest whole number or tenth.

Suggested number of days: 17

		TB: Textbook WB: Workbook	Objectives	Material	Appendix
1.1	**Tenths**				
1.1a	1-Place Decimals	TB: pp. 6-8 WB: pp. 7-9	• Read and write 1-place decimal numbers that are less than 1. • Relate fractions in tenths to 1-place decimal numbers.	• Base-10 blocks • Ruler	• Fraction Squares (pp. a10-a11)
1.1b	1-Place Decimals Greater than 1	TB: pp. 9-10 WB: pp. 10-11	• Read and write 1-place decimal numbers that are greater than 1. • Locate decimal numbers on a number line.	• Base-10 blocks • Ruler • Beaker, graduated cylinder, scale (optional)	• Fraction Squares (pp. a10-a11) • Number Lines - Tenths (p. a12)
1.1c	Decimals and Fractions	TB: p. 10 WB: pp. 12-13	• Express a decimal as a mixed number in simplest form. • Learn some common decimal to fraction conversions. • Compare and order 1-place decimals.	• Base-10 blocks	• Fraction Squares (pp. a10-a11) • Mental Math 1
1.1d	Rename Tenths	TB: p. 11 WB: pp. 14-15	• Rename 10 tenths as 1 one. • Rename 1 one as 10 tenths.	• Place-value discs • Base-10 blocks	• Fraction Squares (pp. a10-a11) • Mental Math 2
1.2	**Hundredths**				
1.2a	2-Place Decimals	TB: pp. 12-15 WB: pp. 16-18	• Read and write 2-place decimal numbers. • Express a fraction with a denominator of 100 as a decimal. • Rename hundredths. • Relate each digit in a 2-place decimal to its place value.	• Base-10 blocks • Place-value discs	• Fraction Squares (pp. a10-a11)

		TB: Textbook WB: Workbook	Objectives	Material	Appendix
1.2b	Hundredths to Decimals	TB: pp. 15-17 WB: pp. 19-22	♦ Express a mixed number as a decimal. ♦ Locate 2-place decimals on a number line. ♦ Relate money in dollars and cents to decimals.	♦ Place-value discs	♦ Number Lines - Hundredths (p. a13)
1.2c	Decimals and Fractions	TB: p. 17 WB: pp. 23-24	♦ Express a 2-place decimal as a fraction. ♦ Express a fraction as a decimal.		♦ Mental Math 3 ♦ Mental Math 4
1.2d	Compare and Order Decimals	TB: pp. 18-19 WB: pp. 25-26	♦ Compare and order numbers to 2 decimal places.	♦ Base-10 blocks ♦ Place-value discs	♦ Fraction Squares (pp. a10-a11)
1.2e	Count On and Count Back	TB: p. 19 WB: pp. 27-28	♦ Add and subtract tenths and hundredths. ♦ Make 1 with hundredths using mental math strategies.	♦ Place-value discs	♦ Mental Math 5 ♦ Mental Math 6
1.3	**Thousandths**				
1.3a	3-Place Decimals	TB: pp. 20-21 WB: pp. 29-30	♦ Read and write 3-place decimal numbers. ♦ Locate 3-place decimals on a number line.	♦ Place-value discs	
1.3b	Compare and Order Decimals	TB: p. 21 WB: p. 31	♦ Compare and order numbers up to 3 decimal places.	♦ Place-value discs	♦ Mental Math 7 ♦ Mental Math 8
1.3c	Decimals to Fractions	TB: p. 22 WB: pp. 32-33	♦ Express a 3-place decimal as a fraction. ♦ Compare and order decimals and fractions.		
1.3d	Practice	TB: pp. 23-24	♦ Practice.		
1.4	**Rounding Off**				
1.4a	Round to a Whole Number	TB: pp. 25-27 WB: pp. 34-35	♦ Round a decimal number to the nearest whole number.		
1.4b	Round to a Tenth	TB: p. 27 WB: p. 36	♦ Round a decimal number to the nearest tenth.	♦ Meter sticks	
1.4c	Review	TB: pp. 28-30 WB: pp. 37-51	♦ Review.		

1.1 Tenths

Objectives

♦ Read and write 1-place decimal numbers.
♦ Convert between fractions (tenths, fifths, or halves) and 1-place decimal numbers.
♦ Locate 1-place decimals on a number line.
♦ Compare and order 1-place decimals.
♦ Rename 10 tenths as 1 one and vice-versa.

Material

♦ Base-10 blocks
♦ Fraction squares (appendix pp. a10-a11)
♦ Number lines (appendix p. a12)
♦ Rulers
♦ Liter beakers or graduated cylinders, if available
♦ Kilogram weighing scale marked in tenths, if available
♦ Place-value discs for 10, 1, and 0.1
♦ Mental Math 1-2

Prerequisites

Students should thoroughly understand the concept of place value. They should also be familiar with fractions, mixed numbers, improper fractions, be able to find equivalent fractions, and convert a fraction to simplest form (lowest terms). If students need more review, revisit equivalent fractions in *Primary Mathematics* 3A and mixed numbers and improper fractions in *Primary Mathematics* 4A. Students should also be familiar with metric measurements and recognize the abbreviations for centimeters, kilograms, and liters.

Notes

In *Primary Mathematics* 2B students were introduced to 2-place decimal numbers in the context of money. Money is written as the number of dollars followed by a decimal point and then the number of cents. So each dollar is a whole and each cent is one hundredth. The decimal point in written money was called a dot.

In this part students will be formally introduced to the tenths place, and learn how to convert between 1-place decimals and fractions.

Place-value notation makes numbers understandable and computation accurate and simple. We use ten digits to write numbers and the actual value of each digit depends on its place. Each digit has a value that is ten times as much as if it were in the place to the right of it and one tenth as much as if it were in the place to the left of it. The number 23,456 represents 2 ten thousands, 3 thousands, 4 hundreds, 5 tens, and 6 ones. The digit 3 is in the thousands place, and its value is 3000. Each whole number can be expanded as the sum of multiples of the value for each place — 1, 10, 100, 1000, etc. So, 23,456 can be written as 20,000 + 3000 + 400 + 50 + 6.

Decimal numbers are an extension of place-value notation to include place values less than 1. We write a decimal point to the right of the ones place. A digit in the first place to the right of the decimal point, the tenths place, has a value that is one tenth of the value of the same digit in the ones place. A digit in the second place to the right of the decimal point, the hundredths place, has a value of one tenth of the same digit in the tenths place. A digit in the third place to the right of the decimal point, the thousandths place, has a value of one tenth of the same digit in the hundredths place.

1.234 is $1 + 0.2 + 0.03 + 0.004$, or $1 + \frac{2}{10} + \frac{3}{100} + \frac{4}{1000}$. We read this number as one and two hundred thirty-four thousandths, or, more simply, one point two three four.

Usually, decimal places are not called by name after the thousandths place; we do not normally say "the hundred-thousandths place" but rather the "fifth decimal place." At this level, students will only encounter decimal numbers to the third place, or thousandths.

If a decimal number is less than 1 we usually use a 0 in the ones place. Writing 0.12 rather than .12 makes it easier to see and pay attention to the decimal point.

Use concrete manipulatives to explain place-value concepts for decimal numbers, including base-10 blocks, fraction squares, number lines, and place-value discs. The example below shows 52.031 with place-value discs on a place-value chart. Place-value discs will be particularly useful in the next unit where they can be used on place-value charts to illustrate addition, subtraction, multiplication, and division. Students should be familiar with place-value discs and charts from earlier levels of *Primary Mathematics*, and should be able to easily extend their use to decimals.

By this level, after using physical discs when a concept is introduced, many students should be able to simply draw circles on an erasable place-value chart to represent numbers and work with renaming concepts such as erasing 10 circles in the tenths place and replacing with a circle in the ones place. More capable students who have a good understanding of place value will not need place-value discs at all by now and the pictorial representation of them in the textbook will be sufficient. Do not force students to use place-value discs at this level if they don't need them. Encourage students who are having difficulties extending place-value concepts to decimals to use or draw the discs.

Tens	Ones	Tenths	Hundredths	Thousandths
10 10 10 10 10	1 1		0.0 0.0 0.01	0.001
5	2	0	3	1

1.1a 1-Place Decimals

Objectives

♦ Read and write 1-place decimal numbers that are less than 1.
♦ Relate fractions in tenths to 1-place decimal numbers.

Vocabulary

♦ Decimals
♦ Decimal point
♦ Tenths place

Note

In the textbook the opening activity consists of looking at measurements, length, weight, and capacity, to illustrate decimal numbers concretely. Decimal numbers are most often used in the context of measurement. Up until now, smaller measurement units have been given a name, e.g., there are 100 centimeters in a meter, so the need for decimals in measurement may not be evident at first. This lesson therefore builds on students' understanding of place value to extend those concepts to decimal numbers before looking at how decimal numbers apply to measurements.

Introduce tenths

Use base-10 blocks or similar material. Draw a place-value chart with three columns with room to add a fourth on the right. Label the columns hundreds, tens, and ones.

Show students the thousand-cube and ask, "What is one-tenth of 1000?" You can also ask them to pick out the correct blocks. Write 100 on the place-value chart. Continue, having students show the block or tell you the value for one tenth of 100 and then one tenth of 10. Write 10 and 1 on the place-value chart.

Show students the unit cube and ask: "What is one tenth of 1?" It is the fraction $\frac{1}{10}$, but because it is one tenth of a 1 in the ones place, we can show it in its own place value, again to the right of the ones. Add a column to the right, label it "Tenths," and write a 1 in it. Tell students that this new place is called the *tenths* place. To show that the digits in this place are less than one whole, we use a dot, called a *decimal point*, to separate it from the whole numbers. We also put a digit in the ones place; for one tenth of 1 we put a 0 in the ones place to show that there are no ones. A number with a decimal point is called a *decimal*. We read the number 0.1 as either "one tenth" or as "zero point one."

Write the number 1111.1. Tell students that each digit has a value that is one tenth of the digit to the left and ten times the digit to the right. The value of the digit is determined by its place. The digit after the decimal point is in the tenths place, and stands for the number of tenths of a whole.

	Hundreds	Tens	Ones
$\frac{1}{10}$ of 1000	1	0	0
$\frac{1}{10}$ of 100		1	0
$\frac{1}{10}$ of 10			1

	Hundreds	Tens	Ones	Tenths
$\frac{1}{10}$ of 1			0	1

$\frac{1}{10}$ of 1 = 0.1

0.1: "one tenth"
       "zero point one"

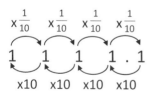

Discussion

Page 6 relates tenths to measurement. Discuss the pictures and ask students what the whole is in each picture (1 cm, 1 kg, 1 ℓ). You can have students look at the markings on their rulers to see that each centimeter is divided into 10 parts. Each picture shows $\frac{8}{10}$ of a whole, which can be written as 0.8. Point out that we read 0.8 as "zero point eight" or as "eight tenths."

Text pp. 6-7

Ask students to say and write the decimal for how much water we would have if we added another tenth to the beaker on p. 6. It would be 0.9 ℓ. Then ask how we would write the decimal if we added another tenth. If we add a tenth to a beaker that already has 0.9 ℓ, we now have 10 tenths of a liter. Since each place can only have the digits 0 through 9, we have to go to the next larger place value, which is one. 1 one is the same as 10 tenths. Write some other tenths and have students "make 1" with them. The top of p. 7 shows a bar divided into tenths, with eight tenths shaded.	0.8 ℓ + 0.1 ℓ = 0.9 ℓ 0.9 ℓ + 0.1 ℓ = 1 ℓ 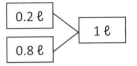
Place value discs and tenths	**Text pp. 7, Task 1**
This task illustrates decimals with place-value discs. Students will work with place-value discs more in the coming lessons. For now, have them write the answers and also say them out loud.	1. (a) 0.4 (b) 0.6 (c) 0.9
Assessment	**Text p. 8, Tasks 2-4**
	2. (a) 0.1 (b) 0.3 (c) 0.5 (d) 0.7 3. 10 4. 4
Practice	WB Exercise 1, pp. 7-9

Activity/Reinforcement

Use or draw fraction squares showing tenths (appendix p. a10). Tell students the square stands for one whole. Color in some rows and have students write the amount colored as both a fraction and a decimal number. For example, color in 3 rows. Students write $\frac{3}{10}$ and 0.3. Continue with some other examples.

Provide students with copies of the fraction squares. Write some 1-place decimal numbers less than 1 and have them color in the correct number of tenths.

$$\frac{3}{10} = 0.3$$

Exercise 1

1. (a) 0.2
   (b) 0.5
   (c) 0.8
   (d) 0.9

2. (a) 0.4 ℓ   (b) 0.7 ℓ

3. (a) 0.9 kg   (b) 0.5 kg

4. (a) 0.2
   (b) 0.6
   (c) 0.9

5. (a) 0.8   (b) 0.4
   (c) 0.5   (d) 0.1

7

8

9

1.1b   1-Place Decimals Greater than 1

Objectives

♦ Read and write 1-place decimal numbers that are greater than 1.
♦ Locate decimal numbers on a number line.

Note

For this lesson use fraction squares showing wholes and tenths, such as those in appendix p. a11, or base-10 blocks. For the fraction squares copy and cut out squares showing wholes and those showing tenths. Students can pick out the appropriate number of wholes and shade in tenths. If they are able to visualize a hundred-flat now as 1 whole, you can use base-10 blocks. Tell them that the flats now represent a whole. You can tape a square sheet of paper on the backside, if that helps. The rods will therefore each represent a tenth. Students can pick out the appropriate number of flats and rods to represent a decimal number.

Introduce tenths greater than one	
Show students some wholes and tenths with fraction squares or with base-10 blocks. Point out that we have some wholes as well as some tenths. Ask them to write the amount first as a fraction, and then as the corresponding decimal number. Then ask them to read the number. The example at the right is read as "two point four" or "two and four tenths." Point out that we use the word "and" between the word for ones and for tenths. 2.4 is four tenths more than 2. Write some other mixed fractions greater than 1 with denominators of 10, and ask students to pick out corresponding flats and rods, or color in some fraction squares, then write them as a decimal number. Then write some improper fractions in tenths and ask them to write them as a decimal number.	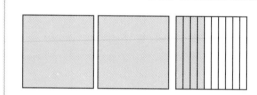 $2\frac{4}{10} = 2.4$ $2.4 = 2 + 0.4$ $7\frac{8}{10} = 7.8$ $\frac{35}{10} = 3.5$
Draw a number line on the board and draw 9 tick marks between each whole number. Ask students to locate specific decimals on the number line. Have students look at the centimeter side of their rulers. Point out the tick marks between the centimeter marks. Each division is a tenth of a centimeter. Ask students to draw some lines correct to the nearest tenth of a centimeter and write the length as a 1-place decimal number. On some rulers the inches from 6 inches to 12 inches are divided into tenths. If the students have rulers with inches divided into tenths, write a 1-place decimal number between 6 and 12 and have them draw a line of the given length in inches. If you have some beakers or graduated cylinders, you can have students measure or read some volumes to the nearest tenth. If you have some weighing scales marked in tenths of a kilogram, you can have them weigh items and write the weights as a decimal to the nearest tenth.	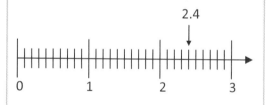

Assessment	Text pp. 9-10, Tasks 5-8

	5. (a) 0.6 cm
	    (b) 0.6
	
	 6. (a) 2.4 ℓ
	    (b) 2.8 kg
	
	 7. (a) 1.5
	    (b) 2.9
	
	 8. A 0.4 B 0.9
	    C 1.1 D 1.6

Practice	WB Exercise 2, pp. 10-11

Activity

Use the number lines in the appendix on p. a12 showing tick marks for tenths or fifths. Write some 1-place decimal numbers (such as 3.2, 6.4, and 22.9) and ask students to locate them on the number lines. If you are working directly with a student or projecting a copy on the board, you can point to a mark and have a student supply the decimal number for that point.

Exercise 2

1. 6.3 cm

2. (a) 6.4 cm
   (b) 9.7 cm
   (c) 8.2 cm

3. (a) 1.6 ℓ
   (b) 2.4 ℓ

4. (a) 2.8 kg
   (b) 1.4 kg

10

11

1.1c   Decimals and Fractions

Objectives

♦ Express a decimal as a mixed number in simplest form.
♦ Learn some common decimal to fraction conversions.
♦ Compare and order 1-place decimals.

Note

Fraction squares, as well as number lines, are used in this lesson to compare numbers. Fraction squares are somewhat more concrete than place-value discs. If students are more advanced, you can use place-value discs on a place-value chart in order to compare two decimal numbers.

"Simplest form" is sometimes called "lowest term" in other math curricula.

Decimals to fractions	
Use fraction squares showing wholes and tenths, or flats and rods from a base-10 set. Write the decimal number 2.4, and ask students to illustrate the number with the fraction squares by picking out two wholes and coloring in tenths on a third square, or with base-10 blocks. Then ask them to write the number as a mixed fraction in tenths. Remind students that for final answers we always want to have the fraction in simplest form. Ask them if this fraction can be simplified and then write the equivalent fractions. Repeat with a few other examples, with and without the fraction squares. Ask them to simplify the fractions whenever possible.	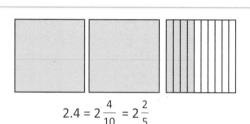 $$2.4 = 2\frac{4}{10} = 2\frac{2}{5}$$ $$3.3 = 3\frac{3}{10} \qquad 10.5 = 10\frac{5}{10} = 10\frac{1}{2}$$
List all the tenths from 0.1 and have students list the equivalent fractions in simplest form. Tell them that these are good equivalencies to memorize.	$0.1 = \frac{1}{10} \qquad 0.2 = \frac{1}{5}$ $0.3 = \frac{3}{10} \qquad 0.4 = \frac{2}{5}$ $0.5 = \frac{1}{2} \qquad 0.6 = \frac{3}{5}$ $0.7 = \frac{7}{10} \qquad 0.8 = \frac{4}{5}$ $0.9 = \frac{9}{10} \qquad 1.0 = 1$
Fractions to decimals	
Write some fractions where the denominators are 2 or 5 and ask students to convert them into decimal numbers. Include some improper fractions. For improper fractions, they need to first convert to a mixed number.	$\frac{4}{5} = \frac{8}{10} = 0.8$ $\frac{3}{2} = 1\frac{1}{2} = 1\frac{5}{10} = 1.5$ $5\frac{2}{5} = 5\frac{4}{10} = 5.4$ $\frac{47}{5} = 9\frac{2}{5} = 9\frac{4}{10} = 9.4$

Compare and order fractions

Refer back to the number line in Task 8, or draw the number line on the board. Ask students which is greater, 1.6 (D) or 0.9 (B) and why. On a number line, the numbers increase from left to right. 1.6 is to the right of 0.9 and is greater than 0.9.

Show the numbers with fraction squares. Write the two numbers vertically, aligning the digits. Point out that we can compare the numbers easily by looking at each digit, starting with the one in the largest place, which in this case is ones. Since 1 one is greater than 0 ones, 1.6 is greater than 0.9.

Repeat with 1.1 (C) and 1.6 (D). 1.6 is again larger. Illustrate with fraction squares. Again, we start by comparing the digit in the highest place (ones). They are the same, so we then compare the digit in the next lower place, which is tenths. Since 6 tenths is greater than one tenth, 1.6 is greater than 1.1.

Write some decimals for students to put in order. You can mix in some mixed numbers with denominators of a factor of 10.

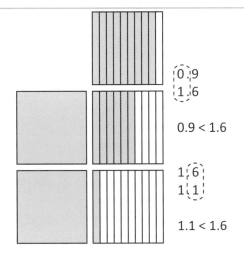

0.9
1.6

0.9 < 1.6

1.6
1.1

1.1 < 1.6

4.2   2.4 $4\frac{2}{5}$ 4

$2.4 < 4 < 4.2 < 4\frac{2}{5}$

Write a number in tenths, and ask students to write a pattern that increases or decreases by a given number of tenths.

3.5; increase by 0.3
3.5, 3.8, 4.1, 4.4, 4.7, 5, 5.3, …

20.1; decrease by 0.5
20.1, 19.6, 19.1, 18.6, 18.1, …

Assessment

Text p. 10, Tasks 9-12

9. (a) $\frac{1}{5}$ (b) $1\frac{1}{5}$

   (c) $\frac{4}{5}$ (d) $2\frac{4}{5}$

10.   3.7
11.   8.5
12.   (a)   0.3, 1.3, 3, 3.1
      (b)   2.7, 7.2, 7.8, 9

Mental Math Practice

Mental Math 1

Practice

WB Exercise 3, pp. 12-13

Exercise 3

1.

0.1	0.2	**0.3**	**0.4**	**0.5**	0.6
$\frac{1}{10}$	**$\frac{2}{10}$**	$\frac{3}{10}$	$\frac{4}{10}$	$\frac{5}{10}$	**$\frac{6}{10}$**

1.1	1.2	**1.3**	**1.4**	2.2	**3.5**
$1\frac{1}{10}$	**$1\frac{2}{10}$**	$1\frac{3}{10}$	$1\frac{4}{10}$	**$2\frac{2}{10}$**	$3\frac{5}{10}$

2. (a) 0.4   (b) 1.4
   (c) 0.5   (d) 3.5

3. (a) $\frac{3}{10}$ (b) $2\frac{3}{10}$

   (c) $\frac{3}{5}$ (d) $3\frac{3}{5}$

4. (a) 0.4; 1.3; 2.8

(b) 8.8; 10.2; 11.7
(c) 59.5; 61.6; 64.4

5. (a) >   (b) >
   (c) =   (d) >

6. (a) 0.1
   (b) 0.9

7. (a) 6.2
   (b) 2.9

8. 2.7, 2.9
   6, 6.5

12   13

1.1d Rename Tenths

Objectives

♦ Rename 10 tenths as 1 one.
♦ Rename 1 one as 10 tenths.

Vocabulary

♦ Whole part
♦ Fractional part

Note

In order to apply the addition and subtraction algorithms to decimals, students will have to rename tenths as ones and tenths, or ones and tenths as tenths. This is no different from what they have already learned with whole numbers and can easily be illustrated with place-value discs in the same way as with whole numbers by trading in a one for 10 tenths or vice-versa. Students should realize that extending the decimal system to numbers less than 1 does not change the basic methods or concepts behind computation.

Renaming	
Use two fraction squares showing tenths. Tape them together and color 14 tenths. Or use 14 rods from a base-10 set. Ask students to write the decimal number for 14 tenths. They should write 1.4. They may write 0.14, which is incorrect. Point out that all tenths need to be in a single place, the tenths place, just like all ones are in a single place, the ones place. Trade in 10 tenths for 1 whole by covering up the 10 tenths with a whole square or replacing 10 rods with a flat. If there are 10 tenths, they must be renamed as 1 one in order to write the amount down with numerals. 14 tenths is renamed as 1 one and 4 tenths. Point out that a number is made up of two parts: the *whole part* and the *fractional part*. In this example, the whole part is 4 and the fractional part is 0.4.	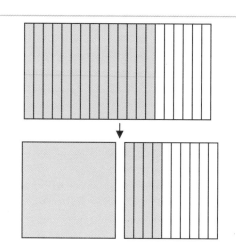 14 tenths = 1 one 4 tenths = 1.4

Ask students the following questions:			
How many tenths are there in 0.4?	(4)	0.4 = 4 tenths	
How many tenths are there in 1?	(10)	1 = 10 tenths	
How many tenths are there in 1.4?	(14)	1.4 = 14 tenths	
How many tenths are there in 4?	(40)	4 = 40 tenths	
What is the decimal number for 42 tenths?	(4.2)	42 tenths = 4.2	

Place-value discs	
Provide students with place-value discs labeled with 10, 1, and 0.1. There should be about 10 tens and 30 ones and tenths per group of students, with students working in pairs or larger groups if there are not enough discs. Mix them up in a bowl or bag for each group. Have them draw a place-value chart with 3 columns labeled *tens*, *ones*, and *tenths*. Then have them each take a	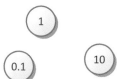

handful of discs, arrange them on the place-value chart, and write the value of the discs. If they have more than 10 of any one kind of place-value disc, they must trade them in for the next higher place value. Ask students to rewrite the decimal number in expanded form. Repeat with other handfuls of discs as needed.	2 (10) 11 (1) 15 (0.1) = 32.5 32.5 = 30 + 2 + 0.5
Assessment	**Text p. 11, Tasks 13-14**

	13. (a) 2.3
	    (b) 36.5
	    (c) 50.4
	
	14. (a) 1.2
	    (b) 2.1

Reinforcement	
Write some equations such as those at the right and have students supply the decimal number. If needed, they can illustrate the problems with place-value discs.	$10 + 0.2 = ?$ 10.2 $0.4 + 4 + 10 = ?$ 14.4 $6 + 0.2 + 30 = ?$ 36.2 $100 + \dfrac{3}{10} + 2 = ?$ 102.3 $30 + ? + 4 = 34.2$ 0.2 $200 + ? + 0.4 = 243.4$ 43
Mental Math Practice	Mental Math 2
Practice	WB Exercise 4, pp. 14-15

Exercise 4

1. (a) 34.6
   (b) 50.7
   (c) 45.3
   (d) 40.9

2. (a) 0.8
   (b) 0.3
   (c) 90
   (d) 30
   (e) 5
   (f) 9

3. See chart at right.

14

15

2.1	1.2	$\dfrac{2}{10}$	$1\dfrac{5}{10}$	5
0.1	$2\dfrac{1}{10}$	$1\dfrac{2}{10}$	0.5	1.5
0.3	$\dfrac{9}{10}$	0.9	$\dfrac{5}{10}$	0.8
$1\dfrac{3}{10}$	4.1	$4\dfrac{1}{10}$	$2\dfrac{8}{10}$	$3\dfrac{7}{10}$
1.3	$\dfrac{4}{10}$	2.8	3.7	6
0.4	1.4	$1\dfrac{4}{10}$	$\dfrac{6}{10}$	0.6

1.2 Hundredths

Objectives

♦ Read and write 2-place decimal numbers.
♦ Locate 2-place decimals on a number line.
♦ Relate money in dollars and cents to decimals.
♦ Express a 2-place decimal as a fraction.
♦ Express specific fractions as decimals.
♦ Compare and order numbers to 2 decimal places.
♦ Add and subtract tenths and hundredths.

Material

♦ Base-10 blocks
♦ Fraction squares (appendix pp. a10-a11)
♦ Number lines (appendix p. a13)
♦ Place-value discs marked with 10, 1, 0.1, and 0.01
♦ Mental Math 3-6

Prerequisites

In addition to the prerequisites for the previous part, students should be familiar with mental math strategies which involve adding a single digit to a multi-digit number or subtracting a single digit from a multi-digit number, even when there is renaming, such as with $458 + 9$ or $203 − 6$. They should also be able to mentally "make 100" in order to subtract from 100, as in $100 − 42 =$ ___ or the related problem $42 +$ ___ $= 100$. If they are not comfortable with these mental math strategies, revisit the mental math strategies in the Teacher's Guide for *Primary Mathematics* 2A, Unit 2 and *Primary Mathematics* 2B, Unit 3.

Notes

In this part decimal notation is extended to 2 decimal places. The first decimal place is the tenths place and the second is the **hundredths place**. The value of the digit in the hundredths place is one tenth of what that digit would be in the tenths place, or one hundredth of what it would be in the ones place.

$$0.04 = \frac{4}{100} \text{ of 1 whole}$$

$$0.73 = \frac{7}{10} + \frac{3}{100} = \frac{70}{100} + \frac{3}{100} = \frac{73}{100} \text{ of 1 whole}$$

Students will be converting decimals to fractions in simplest form. Equivalent fractions and fractions in simplest form were covered in *Primary Mathematics* 3A and 4A and factors were covered in *Primary Mathematics* 4A. At this level students will only be converting fractions which, in their simplest form, have denominators that are factors of 100 to 1-place or 2-place decimal numbers by simply finding an equivalent fraction with a denominator of 10 or 100. For example:

$$\frac{3}{25} = \frac{12}{100} = 0.12$$

Students will learn how to use division to convert all fractions to decimals in *Primary Mathematics* 5A.

In some other school curricula students are taught to only read decimal numbers as fractions, using the word "and" for the decimal point. 6.58 is read as "six and fifty-eight hundredths" and 2.248 is read as "two and two hundred forty-eight thousandths." You should teach students to recognize and use this nomenclature through thousandths, as they may encounter it on standardized tests. However, this nomenclature is cumbersome for decimals with place values larger than hundredths, and is not normally used in algebra and more advanced mathematics. In the *Primary Mathematics* curriculum students are allowed to use the nomenclature where the decimal point is read as "point" and then the digits simply read. 2.248 is "two point two four eight." If students already understand place value, which they should by now, this nomenclature should not cause any difficulties with understanding what each digit in each place stands for.

Decimal numbers are compared to see which is larger in the same way as whole numbers. We start by comparing the digits in the highest place value. If they are the same, we then compare the digits in the next higher place value, and so on. As with whole numbers, we have to be careful to pay attention to the place value of the digits, not the number of digits. 4.5 is larger than 4.25 even though it has fewer digits.

In this part students will be adding or subtracting a number with a small non-zero digit to or from a decimal number. For example, 40.92 + 0.2 or 5.61 − 0.03. These can be done by counting up or down. The purpose is to pay attention to the place value of the digit being added or subtracted. With larger non-zero digits students should be able to extend mental math strategies learned earlier for adding a single non-zero digit to tenths and hundredths. For example, if 68 ones + 5 ones = 73 ones, then 68 hundredths + 5 hundredths = 73 hundredths (as in 5.68 + 0.05 = 5.73) and 68 tenths + 5 tenths = 73 tenths = 7.3 (as in 6.8 + 0.5 = 7.3 or 6.81 + 0.5 = 7.31). If students cannot do the problems mentally, you may want to review mental math strategies from *Primary Mathematics* 3A and 4A; adding a single non-zero digit mentally will not be re-taught here. If students have not learned mental math strategies yet, you can have them answer the problems in the workbook by simply counting up or down. You will have to take time to teach them from earlier levels, and in the meantime skip the last lesson in this part. Students will formally learn to add and subtract decimal numbers using the standard algorithm in the next unit, which they can also use until they learn the mental math strategies.

In *Primary Mathematics* 2B students learned how to "make 100." This was extended to making change for a dollar. Students can use the same strategies to "make 1" or "make the next 1" with a 2-place decimal. For example, since 47 must be added to 53 to make 100, 0.47 must be added to 0.53 to make 1.

1.2a   2-Place Decimals

Objectives

♦ Read and write 2-place decimal numbers.
♦ Express a fraction with a denominator of 100 as a decimal.
♦ Rename hundredths.
♦ Relate each digit in a 2-place decimal to its place value.

Vocabulary

♦ Hundredths place
♦ Decimal places

Note

Fraction squares again provide a good concrete representation of decimal numbers. You can also use base-10 blocks, in which case the flats are ones, the rods are tenths, and the unit cubes are hundredths.

Introduce hundredths

Display a fraction square divided into ten columns and ask students what each column represents. Now show one where each column is divided into ten equal parts. Color in one square. Ask students to compare it to the fraction square showing tenths and tell you what fraction of a tenth and of the whole it is. The colored square is 1 tenth of 1 tenth, and also 1 hundredth of 1 whole.

Write 1 on a place-value chart with a column for ones, add a column for tenths and write 0.1, and then ask students what we need to do to show a tenth of a tenth. Add a column for hundredths and write 0.01. Tell students that this new place is called the *hundredths place*.

$\frac{1}{10}$ of 1 = 0.1

$\frac{1}{10}$ of 0.1 = 0.01

Ones	Tenths	Hundredths
1		
0	1	
0	0	1

Discussion

The picture at the top of p. 12 in the text shows a bar representing 1 whole divided into tenths, and then one of tenths further divided into 10 parts, which is expanded. Ask students how many such small parts would be in the entire bar for 1. There would be 100 small parts. Each small part is a tenth of a tenth. Three tenths and seven hundredths are shaded, or 0.37. Point out the two methods for reading a decimal. Be sure students understand why 0.37 = 37 hundredths = 3 tenths 7 hundredths. 0.37 is 30 hundredths and 7 hundredths. From the picture they can see that each 10 hundredths is the same as 1 tenth.

Text pp. 12-13

0.37 = 37 hundredths
   = 3 tenths 7 hundredths
   = 30 hundredths + 7 hundredths

Place-value discs

Task 1 uses place-value discs to show that 1 tenth can be renamed as 10 hundredths. For (c) point out that we can replace the 10 0.01-discs with a 0.1-disc.

Display nine 0.1-discs and ten 0.01-discs. Ask students to write the number these discs represent. We have to rename the 10 hundredths as 1 tenth, and then the 10 tenths as 1 one, in order to write the number represented by the discs.

Text p. 13, Task 1

1. (a) 0.03   (b) 0.05   (c) 0.12

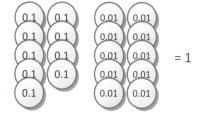

Point out that adding 0's on after the decimal point does not change the value of the number. So 1 is the same as 1.0 is the same as 1.00.	1 = 1.0 = 1.00 1 = 10 tenths = 100 hundredths

Discussion	**Text pp. 14-15, Tasks 2-4**

Task 2: Ask students to also write both the decimal number and the fraction or mixed number with 100 in the denominator for each problem. For Task 2(a) point out that the 0 in the tenths place means that there are no tenths. Remind them that each 10 hundredths is the same as 1 tenth. So $4\frac{25}{100}$ is the same as $4 + \frac{2}{10} + \frac{5}{100}$.
 	
Task 3: In this task the discs are shown on a place-value chart. You can expand on this task by displaying other numbers with place-value discs on a chart and asking students what each digit stands for. Include some where there are no discs in a column so that 0 needs to be used.
 	
Task 4: In this task the discs on the chart are replaced with just numbers.
	2. (a) $3.02, 3\frac{2}{100}$
	
	   (b) $4.25, 4\frac{25}{100}$
	
	3. 3: 30
	   4: 4
	   5: 0.5 or $\frac{5}{10}$
	   6: 0.06 or $\frac{6}{100}$
	4. 9: 0.9 or $\frac{9}{10}$
	   2: 0.02 or $\frac{2}{100}$
	   7: 7
	   4: 40
	   3: 300

Assessment	
Write down a number without a place-value chart and ask students for the value of each digit. See the example at the right.	205.04: What digit is in the tenths place?(0) What is the value of the digit 2? (200) What is the value of the digit 4? (4 hundredths)

Practice	**WB Exercise 5, pp. 16-18**

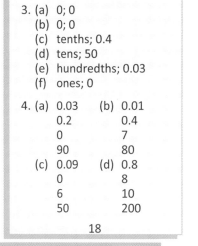

Exercise 5

1. (a) 0.82
   (b) 8.34
   (c) 3.05
   (d) 5.17
   (e) 20.09

16

2. (a) 34.02
   (b) 40.25
   (c) 24.13
   (d) 30.04

17

3. (a) 0; 0
   (b) 0; 0
   (c) tenths; 0.4
   (d) tens; 50
   (e) hundredths; 0.03
   (f) ones; 0

4. (a) 0.03   (b) 0.01
       0.2   0.4
       0   7
       90   80
   (c) 0.09   (d) 0.8
       0   8
       6   10
       50   200

18

1.2b   Hundredths to Decimals

Objectives

- Express a mixed number as a decimal.
- Locate 2-place decimals on a number line.
- Relate money in dollars and cents to decimals.

Note

In this lesson students will be looking at number patterns. In a number pattern such as the following: 0.08, 0.1, 0.12, two hundredths are added each time. If students become confused because of the change from 2 digits after the decimal point to 1 digit after the decimal point back to 2 digits, show the first number with fraction squares or place-value discs. Ask them by how much the next number increases, and then color in two more tenths or add more place-value discs, renaming as needed. Students should be able to see that we are simply going from 10 hundredths to 1 tenth and do not write the 0 after the 0.1. You can add in the 0 if that helps.

Number lines	
Use the number lines on appendix p. a13. Write some decimals on the board for students to locate and mark on the number lines. Students should realize there are only 5 divisions on the second and third number line. A decimal with an odd number in the hundredths place would therefore be located between two tick marks.	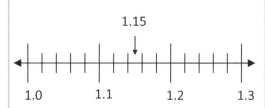
Discussion	**Text p. 15, Task 5**

5(a): This task shows that 40 hundredths is the same as 4 tenths. Each little square is a hundredth, each column of 10 little squares is a tenth, so 40 little squares is the same as 4 tenths. Point out that we can write $\frac{40}{100}$ as 0.40, but this is the same as 0.4. Remind students that it does not matter if we add 0's on the end after the last non-zero decimal number. 4 tenths is the same as 40 hundredths.	5. (a) 0.4
	   (b) 1.28
	   (c) 2.05
	
	0.4 = 0.40
	4 tenths = 40 hundredths

5(b): Point out that in the second square two columns are colored. You can break out the fractional part as the sum of tenths and hundredths.	$\frac{28}{100} = \frac{20}{100} + \frac{8}{100} = \frac{2}{10} + \frac{8}{100} = 0.2 + 0.08$

5(c): Make sure students understand that other than the two wholes, there are no tenths, so the decimal number has a 0 in the tenths place. A 0 anywhere before the last non-zero digit after the decimal does change the number. 2.5 is not the same as 2.05.	2.5 = 2 wholes 5 tenths
	   = 2.50
	   = 2 wholes 50 hundredths
	2.05 does not equal 2.5
	2.05 = 2 wholes 5 hundredths

Number patterns	
Write the following number patterns on the board and discuss them with the students. If necessary append 0's so they can see how the hundredths are increasing, e.g., 0.25, 0.50, 0.75, 1.00. Use fraction squares or place-value discs if needed.	0.04, 0.06, 0.08, _____, _____ 0.1, 0.12 0.25, 0.5, 0.75, _____, _____ 1, 1.25 0.07, 0.1, 0.13, _____, _____ 0.16, 0.19 3, 3.05, 3.1, 3.15, _____, _____ 3.2, 3.25

Decimals and money	Text p. 16, Task 6

6(a): Remind students that we write money as the number of dollars, a dot, and the number of cents, allowing two places for cents. The dot is the same as a decimal point; the dollar is the whole. There are 100 cents in a dollar, so each cent is $\frac{1}{100}$ of a dollar, or $0.01.
 	
6(b): 10 cents is the same as $\frac{10}{100}$ of a dollar, or $0.10. Point out that 10 cents is also a tenth of a dollar, but with money the two decimal places represent the number of cents, and so we always use both decimal places after the dot; we do not write 10 cents as $0.1.
 	
Have students supply answers to 6(c)-6(f).
	6. (a) $\frac{1}{100}$ (b) $\frac{1}{10}$
	
	(c) $\frac{2}{10}$ (d) $\frac{5}{10}$
	   20¢ $0.50
	
	(e) $\frac{45}{100}$ (f) $\frac{26}{100}$
	   45¢ $0.26

Assessment	**Text pp. 16-17, Tasks 7-10**

	7. (a) $3.85 (b) $6.50
	  (c) $8.05 (d) $85.00
	
	8. (a) 2.84 (b) 36.25
	  (c) 54.03 (d) 80.57
	
	10. (a) A: 0.04 B: 0.07 C: 0.11
	         D: 0.13 E: 0.19
	    (b) P: 4.62 Q: 4.66 R: 4.69
	         S: 4.73 T: 4.78

Write the following number patterns on the board and have students complete them.	2, 2.04, 2.08, ____, ____, 2.2 2.12, 2.16 10, 9.8, 9.6, ____, ____, ___ 9.4, 9.2, 9 1.65, 1.6, ____, ____, 1.45 1.55, 1.5
Practice	WB Exercise 6, pp. 19-20 WB Exercise 7, pp. 21-22

Exercise 6

1. (a) 0.07   (b) 1.07
  (c) 0.58   (d) 2.58
  (e) 0.24   (f) 1.24
  (g) 0.65   (h) 3.65
  (i) 0.03   (j) 2.03
  (k) 0.05   (l) 10.05

19

2. $\frac{9}{10} \rightarrow 0.9$

$\frac{17}{100} \rightarrow 0.17$

$\frac{7}{100} \rightarrow 0.07$

$\frac{3}{10} \rightarrow 0.3$

$\frac{29}{100} \rightarrow 0.29$

$\frac{7}{10} \rightarrow 0.7$

$\frac{9}{100} \rightarrow 0.09$

20

Exercise 7

1. (a) 80.7   (b) 24.5
  (c) 34.04   (d) 7.29

2. (a) $\frac{7}{100}$ (b) $\frac{5}{100}$

  (c) $\frac{2}{10}$ (d) $\frac{7}{10}$

  (e) $\frac{4}{10}$

3. (a) 0.04   (b) 0.05
  (c) 0.1   (d) 0.08
  (e) 0.3

21

4. (a) 1; 1.2
  (b) 3, 3.5
  (c) 2.7; 2.5
  (d) 8.5; 7.5
  (e) 0.2; 0.3
  (f) 0.3, 0.25
  (g) 0.08; 0.12
  (h) 9.85; 9.75

5. (a) 0.13; 0.28
  (b) 0.87; 0.97
  (c) 3.08; 3.22; 3.37

22

1.2c   Decimals and Fractions

Objectives

- Express a 2-place decimal as a fraction in simplest form.
- Express a fraction with a denominator that is a factor of a 100 as a decimal.

Note

In this lesson students will only convert fractions to decimals where the decimal has a denominator that is a factor of 100. They simply have to find the equivalent fraction with a denominator of 100, then write the decimal. By first converting decimals to fractions and then simplifying, they can become familiar with fractions that can then easily be converted back into a decimal. Although fractions can be converted into decimals using division, students will not look at the relationship between fractions and decimals until *Primary Mathematics* 5.

Review	
Write some equations such as those shown on the right on the board and have students supply the answer as a decimal number.	$0.44 + 4 + 40 = ?$ 44.44
	$6 + 0.2 + 30 + \dfrac{1}{100} = ?$ 36.21
	$100 + \dfrac{3}{10} + 0.02 + 8 = ?$ 108.32
	$\dfrac{34}{100} + 100 = ?$ 100.34
	450 tenths = ? 45
	450 hundredths = ? 4.5
	3.33 = ? hundredths 333
Discussion	**Text p. 17, Task 10**
Task 10(a): Ask students to simplify this fraction. Then ask them what coin could be $0.25. (A quarter). Since there are four quarters in a dollar, then one quarter is a fourth of a dollar. Ask students to use this idea to find 0.75 as a fraction without first converting it to a fraction with a denominator of 100: $0.75 = 3 quarters = $\dfrac{3}{4}$ of a dollar.	10. (a) $\dfrac{1}{4}$ (b) $1\dfrac{21}{25}$ $0.75 = 3$ quarters of $100 = \dfrac{3}{4}$
Assessment	**Text p. 17, Task 11**
	11. (a) $\dfrac{3}{50}$ (b) $\dfrac{7}{25}$ (c) $\dfrac{6}{25}$ (d) $2\dfrac{1}{20}$ (e) $3\dfrac{13}{20}$ (f) $4\dfrac{3}{4}$
Discussion	
After students have finished Task 11 and have checked their answers, ask them to find the factors of 100 and compare them to the denominators in the answers. All 1-place or 2-place decimal numbers rewritten as fractions in their simplest form will have only those numbers as denominators.	Factors of 100: 1, 2, 4, 5, 10, 20, 25, 50, and 100

Ask students which decimal numbers will have 100 as the denominator of the fraction in simplest form. These will be any decimal that does not have a common factor with 100, such as 0.49. Ask students which 2-place decimal numbers might have 20 as the denominator of the simplified fractions. These would be any decimals with a 5 in the hundredths place, e.g., 0.45.	
Discussion	**Text p. 17, Task 12**
Task 12: This task illustrates that we have to first convert the fraction to an equivalent fraction with 10 or 100 in the denominator in order to express it as a decimal.	12. (a) $\frac{6}{10}$ = **0.6** (b) $\frac{45}{100}$ = **0.45**
Assessment	**Text p. 17, Task 13**
	13. (a) 0.75 (b) 0.35 (c) 0.32 (d) 1.5 (e) 2.4 (f) 3.54
Discussion	
After students have finished Task 13 and checked their answers, list the fractions at the right and have them express each as a decimal. Tell them that if they memorize these 8 equivalencies, it will be easy to convert them and their multiples to decimals. If you have capable students you can give some examples now, but students will learn to multiply a decimal by a whole number in the next unit.	$\frac{1}{2}$ = 0.5 $\frac{1}{50}$ = 0.02 $\frac{1}{4}$ = 0.25 $\frac{1}{25}$ = 0.04 $\frac{1}{5}$ = 0.2 $\frac{1}{20}$ = 0.05 $\frac{1}{10}$ = 0.1 $\frac{1}{100}$ = 0.01 If $\frac{1}{5}$ = 0.2, then $\frac{3}{5}$ = 3 x 0.2 = 0.6 If $\frac{1}{25}$ = 0.4, then $\frac{4}{25}$ = 4 x 0.04 = 0.16
Mental Math Practice	Mental Math 3-4
Practice	WB Exercise 8, pp. 23-24

Exercise 8

1. (a) $\frac{1}{2}$ (b) $2\frac{1}{2}$

   (c) $\frac{2}{25}$ (d) $1\frac{2}{25}$

   (e) $\frac{3}{20}$ (f) $3\frac{3}{20}$

   (g) $\frac{16}{25}$ (h) $1\frac{16}{25}$

2. $\frac{2}{10}$, 0.2

3. $\frac{75}{100}$, 0.75

23

4. (a) $\frac{5}{10}$ = 0.5   (b) $3\frac{5}{10}$ = 3.5

   (c) $\frac{6}{10}$ = 0.6   (d) $1\frac{6}{10}$ = 1.6

   (e) $\frac{25}{100}$ = 0.25 (f) $2\frac{25}{100}$ = 2.25

   (g) $\frac{16}{100}$ = 0.16 (h) $1\frac{16}{100}$ = 1.16

5. (a) 0.8   (b) 3.8
   (c) 0.45   (d) 1.45
   (e) 0.06   (f) 2.06

24

1.2d   Compare and Order Decimals

Objectives

♦ Compare and order numbers to 2 decimal places.

Note

Comparing decimals follows the same process as comparing whole numbers; the digits in the highest place value are compared first, and if they are the same then the digits in the next lower place value are compared. Students still need to visualize what is happening, so use fraction squares or base-10 blocks as needed.

Compare decimals	
Provide students with fraction squares or base-10 blocks. Write the two numbers 1.01 and 1.1 on the board and ask students to use the fraction squares or blocks to show the numbers. Then ask them which one is larger and discuss why it is larger. 1.1 is larger than 1.01 because a tenth is larger than a hundredth. Point out that we cannot compare the numbers by simply looking at the number of digits. Although 11 is smaller than 101 because it has no hundreds, 1.1 is not smaller than 1.01. Repeat with 1.2 and 1.22. This time 1.22 is larger because it has two more hundredths. Repeat with other examples as needed.	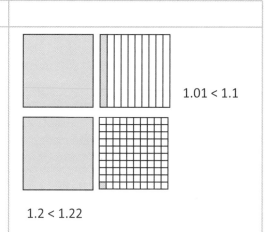 1.01 < 1.1 1.2 < 1.22

Discussion	Text p. 18, Tasks 14-15

Task 14(a): Make sure students realize that even though 2.9 has fewer digits, it is larger. The ones are the same, so we compare the tenths. 9 tenths (in 2.9) is greater than 1 tenth (in 2.12). Ask students to write an inequality.	14. (a) 2.9 is greater. 2.9 > 2.12
	   (b) 1.68 is smaller. 1.68 < 2.35
	
	15. (a) 562.41
	   (b) 89.67

Task 14(b): All we need to compare is the ones. Remind students that we start with the digits in the highest place in order to compare numbers.	
Task 15(a): Ask students to explain why 562.41 is greater than 562.38. Hundreds, tens, and ones of both numbers are the same, but 4 tenths is greater than 3 tenths, so 562.41 is greater.	
Task 15(b): Again, ask students to explain their answer. Make sure they understand that even though both numbers have 4 digits and 243.5 starts with a smaller digit than 89.67, the number is larger because it has hundreds and 89.67 has no hundreds.	

Assessment	Text p. 19, Tasks 16-17

If necessary students can rewrite the problems one on top of the other, or on a place-value chart, aligning the digits according to place value.	16. (a) 42.6
	    (b) 2.5 m
	    (c) 32.6 kg
	
	17. (a) 2.2, 2.02, 0.2, 0.02
	    (b) 80.7, 74.5, 7.8, 7.45

Reinforcement	
Write the numbers on the right on the board and have students put them in ascending order.	10.02, 10.25, 10.2, 12.5, 12.05, 1.25 Answer: 1.25, 10.02, 10.2, 10.25, 12.05, 12.5
Practice	WB Exercise 9, pp. 25-26

Group Game

Material: Place-value discs in a bag (100's, 10's, 1's, 0.1's, and 0.01's) for each group.

Procedure: Each player draws 10 discs then writes down the number formed. The player with the greatest number gets a point. The winner is the one who gets 10 points (or some other target number) first.

Class Game

Material: Numbers of up to 2 decimal places written on index cards. Include whole numbers, tenths, and hundredths, as many cards as students.

Procedure: Divide the class up into 2 teams. Give each student a card. Students must line up in order according to the numbers on their cards. The first team to be in order wins.

Exercise 9

1. (a) >

   (b) >

   (c) <

   (d) >

25

2. (a) <   (b) >
   (c) <   (d) >
   (e) =   (f) >
   (g) =   (h) >

3. (a) 0.88
   (b) 0.61
   (c) 2.99
   (d) 0.42

4. (a) 3
   (b) 8.1
   (c) 5.33
   (d) 7.01

26

Objectives

- Add and subtract tenths and hundredths to decimals up to 2 decimal places.
- Make 1 with hundredths using mental math strategies.

Note

Although students can do the problems in this lesson by counting on or back, those who thoroughly understand place value and are comfortable with mental math should not have trouble extending mental math strategies to decimal numbers. For slower students display the numbers with place-value discs on a place-value chart and show what happens when adding or subtracting one by one to the digit in a specific place.

Count up or down	
Write the expression **14.88 + 0.01** on the board. Ask students for the answer. Point out that we are adding a hundredth, so we increase the digit in the hundredths place. Illustrate with place-value discs if needed. Then write the expression **14.89 + 0.1**. Again, we are adding hundredths, but this time we will have ten hundredths so we rename them as a tenth. In a similar fashion discuss the other problems shown at the right. In order to count down or take away 0.1 or 0.01 if there is a 0 (or no digit) in the tenths or hundredths place, a one or a tenth has to be renamed. Draw attention to the place value to which we are adding to or subtracting from. For example, in 14.19 − 0.1, students should notice that we are subtracting tenths, not hundredths.	$14.88 + 0.01 = ?$ 14.89 $14.89 + 0.01 = ?$ 14.9 $14.9 + 0.01 = ?$ 14.91 $14.91 + 0.1 = ?$ 15.01 $14.21 − 0.01 = ?$ 14.2 $14.2 − 0.01 = ?$ 14.19 $14.19 − 0.1 = ?$ 14.09 $14.09 − 0.1 = ?$ 13.99 $13 − 0.01 = ?$ 12.99
Have students add or subtract 0.2, 0.02, 0.3, and 0.03 from some decimals by counting up or down.	$9.47 + 0.02$ 9.49 $9.47 + 0.03$ 9.5 $4.82 + 0.2$ 5.02 $2.22 − 0.2$ 2.02 $2.22 − 0.3$ 1.92
Make a 1	
Write the equation **48 + ___ = 100** on the board and discuss the strategies that can be used to find the missing number. 100 is the same as 9 tens and 10 ones, so we can easily subtract 4 tens from 9 tens and 8 ones from 10 ones. Write the equation **0.48 + ___ = 1**. Since 0.48 is 48 hundredths, and 1 is 100 hundredths, we can use a similar strategy to find the answer.	$48 + ___ = 100 \rightarrow 100 − 48$ 9 tens 10 ones − 4 tens 8 ones = 52 $0.48 + ___ = 1 \rightarrow 1.00 − 0.48$ $1 \Big\langle \begin{matrix} 0.9 − 0.4 = 0.5 \\ 0.10 − 0.08 = 0.02 \end{matrix} \Big\rangle 0.52$

Assessment	Text p. 19, Tasks 18-23

	18. (a) 412.44
	    (b) 412.24
	19. (a) 123.49
	    (b) 123.47
	20. (a) 5.17
	    (b) 28.6
	21. (a) 86.63 (b) 24.85 (c) 4.89
	    (d) 54.22 (e) 6.2 (f) 3.43
	22. (a) 0.54 (b) 0.04
	23. 0.18

Mental math	

Since the exercise contains problems involving adding and subtracting digits larger than 3, you may want to review mental math strategies and show students how they can apply them to decimal numbers.
 	
Discuss strategies students have learned to add 58 and 6, such as making the next ten. Similarly, we can add 0.58 and 0.06 and 5.8 and 0.6 or 5.82 and 0.6 using the same strategy but with hundredths or tenths, essentially making the tenth or next whole. Point out that in 5.82 + 0.6 we are adding to the tenths; we are not adding hundredths so the 2 hundredths does not change.
 	
Similarly, discuss strategies for subtracting 7 from 52 (such as subtracting from the ten), and then 7 hundredths from 52 hundredths (0.52 – 0.7) or 7 tenths from 52 tenths (5.2 – 0.7).
	$58 + 6 = 60 + 4 = 64$
	     ∧
	   2 4
	$0.58 + 0.06 = 0.60 + 0.04 = 0.64$
	$5.8 + 0.6 = 6.0 + 0.4 = 6.4$
	$5.82 + 0.6 = 6.02 + 0.4 = 6.42$
	
	$52 - 7 = 43 + 2 = 45$
	   ∧
	50 2
	$0.52 - 0.07 = 0.43 + 0.02 = 0.45$
	$5.2 - 0.7 = 4.3 + 0.2 = 4.5$
	$5.22 - 0.7 = 4.32 + 0.2 = 4.52$

Enrichment - Mental Math Practice	Mental Math 5-6
Practice	WB Exercise 10, pp. 27-28

Exercise 10

1. (a) 324.57
   (b) 234.05

2. (a) 46.15
   (b) 39.21
   (c) 59.98
   (d) 42.49
   (e) 0.1
   (f) 0.01
   (g) 0.1
   (h) 0.01

27

3. (a) 5.56   (b) 4.95
   (c) 4.02   (d) 7.23
   (e) 4.58   (f) 8.1
   (g) 6.5   (h) 5.34

4. (a) 2.33   (b) 4.68
   (c) 3.98   (d) 1.64
   (e) 3.45   (f) 4.22
   (g) 5.19   (h) 3.63

5. (a) 0.55   (b) 0.38
   (c) 0.99   (d) 0.92

28

1.3 Thousandths

Objectives

♦ Read and write decimal numbers for thousandths.
♦ Compare and order decimal numbers.
♦ Write a decimal number as a fraction in simplest form.
♦ Compare and order a mixture of whole numbers, decimals, and fractions.
♦ Count up or down by hundredths or thousandths.

Material

♦ Place-value discs for 0.001, 0.01, 0.1, 1, 10, and 100
♦ Mental Math 7-8

Notes

In this part decimal notation is extended to 3 decimal places. The third decimal place is the thousandths place. 1 one = 1000 thousandths, 1 tenth = 100 thousandths, and 1 hundredth = 10 thousandths.

$$0.125 = \frac{125}{1000} \text{ of 1 whole}$$

$$0.125 = 0.1 + 0.02 + 0.005 = \frac{1}{10} + \frac{2}{100} + \frac{5}{1000}$$

Students will convert 3-place decimal numbers to fractions, using the same strategies they used with 1-place and 2-place decimal numbers as well as strategies they have already learned for finding equivalent fractions. The resulting fractions will have denominators which are factors of 1000: 1, 2, 4, 5, 8, 10, 20, 25, 40, 50, 100, 125, 200, 250, 500, and 1000.

Students will not be required to convert fractions to decimals that result in 3-place decimals. All the fractions they have to convert at this level will result in decimal numbers of no more than 2 places.

Students will also compare and order decimal numbers of up to 3 decimal places and fractions. To compare fractions and decimals they will generally need to convert the fractions to decimals.

Appending 0's to a decimal after the last digit does not change the value. However, for your information, extra zeroes are written after a decimal number to indicate the precision of a measurement. For example, a measurement of 0.020 m means that the length was measured to the nearest millimeter; it is between 0.0195 m and 0.0204 m. 0.02 m means that the length was measured to the nearest centimeter; it is between 0.015 m and 0.024 m.

Blank Page

1.3a   3-Place Decimals

Objectives

♦ Read and write 3-place decimal numbers.
♦ Locate 3-place decimals on a number line.

Vocabulary

♦ Thousandths place

Note

To illustrate a thousandth, if you have been using fraction squares, show one divided into hundredths, divide one of those up into ten parts, and shade a part. If you have been using base-10 blocks, show the flat for one whole, then the rod for a tenth, then the unit cube for a hundredth, and then tell them to imagine the unit cube sliced into 10 flats; one tiny flat would be one thousandth of the big flat. You can also point out that one of the divisions between the centimeters on a ruler is a millimeter, which is one thousandth of a *meter*. (Previously it was used to illustrate one tenth of a *centimeter*.)

Introduce thousandths	
Ask students to write the number (not a fraction) for one tenth of 10, and name the number, then for one tenth of that, and name the number, then for one tenth of that, and name the number. They may be able to write and name one thousandth on their own. Tell them that in order to write a number for a tenth of a hundredth, we add another place to the right. This is called the *thousandths place* because there are 1000 thousandths in one whole. Then have students write the fractions for one tenth, one hundredth, and one thousandth. Tell students that we could keep on adding new places after the decimal, each one a tenth of the previous one. The next two places would then be the ten-thousandths and the hundred-thousandths places. You may want to discuss why we even need such small parts of a whole. One area where they are used is in measurement. A centimeter is one hundredth of a meter, so a centimeter is the same as 0.01 meters. A millimeter is one thousandth of a meter, and is the same as 0.001 meters. Scientists measure things even smaller than a millimeter. A dust mite is 400 microns long. A micron is one thousandth of a millimeter, and can be written as 0.000001 m, so a dust mite is 0.0004 m long. Writing the measurements as decimals makes it easier to compare sizes and to convert between measures.	10 ten 1 one 0.1 one tenth $\frac{1}{10}$ 0.01 one hundredth $\frac{1}{100}$ 0.001 one thousandth $\frac{1}{1000}$ 1 centimeter = 0.01 m 1 millimeter = 0.001 m 1 micron = 0.000001 m
Discussion	**Text p. 20**
You can show the discs as shown in the textbook on a place-value chart if students have difficulties with writing the decimals. Show other examples as needed, particularly ones where there are no tenths or hundredths.	(a) 0.024 (b) 0.315 (c) 4.002
Show students a 0.001-disc and ask them how many of these are in a 0.01-disc (10), a 0.1-disc (100), and a 1-disc (1000). Place or draw 15 of the 0.001-discs on the chart and ask students to write the decimal number. They must rename 10 thousandths as 1 hundredth.	10 thousandths =1 hundredth 100 thousandths = 1 tenth 1000 thousandths = 1 one 15 thousandths = 1 hundredth + 5 thousandths = 0.01 + 0.005 = 0.015

Then show or draw twelve 0.001-discs and twelve 0.01-discs and ask students to write the decimal number.	12 hundredths + 12 thousandths = 1 tenth + 2 hundredths + 1 hundredth + 2 thousandths = 0.132
Write the equations shown on the right and discuss. The numerator for each fraction is the number of thousandths, so twenty thousandths is 0.020, but that is the same as 0.02. The number of zeros after the last non-zero digit in a decimal does not affect the value of the decimal number.	$\frac{2}{1000} = 0.002$ $\frac{20}{1000} = 0.020 = 0.02$ $\frac{200}{1000} = 0.200 = 0.2$ $\frac{2000}{1000} = 2.000 = 2$

Assessment	**Text pp. 20-21, Tasks 1-3**

	1. (a) five thousandths, 0.005
	   (b) 2: 20, 2 tens
	       0: 0, 0 ones
	       4: 0.4, 4 tenths
	       3: 0.03, 3 hundredths
	
	2. (a) 5.63
	   (b) 5.61
	   (c) 4.537
	   (d) 4.535
	
	3. (a) 0.148
	   (b) 0.048
	   (c) 0.008

In Task 2 students have to pay attention to the place value of the digit. You can expand on this task by using the problems at the right or similar problems. Illustrate with place-value discs if needed, or guide students in counting up or down. For example, with 5.788 + 0.03 we are adding to the hundredths place. 79, 80, 81 hundredths: 5.818.	5.788 + 0.001 5.789 5.788 + 0.002 5.79 5.788 + 0.02 5.808 5.788 + 0.03 5.818 5.111 − 0.001 5.11 5.111 − 0.002 5.109 5.111 − 0.02 5.091

Practice	WB Exercise 11, pp. 29-30

Exercise 11

1. (a) 0.004
   (b) 4.007
   (c) 0.083
   (d) 0.435

2. (a) 0.003
   (b) 0.406

3. (a) $\frac{9}{1000}$

   (b) $\frac{43}{1000}$

29

4. (a) 3 ones, 4 tenths,
       7 hundredths,
       9 thousandths
   (b) 4; 0.4
   (c) $\frac{9}{1000}$ or 0.009
   (d) $\frac{7}{100}$ or 0.07

5. (a) 8.4, 8.8, 9.1, 9.5
   (b) 3.22, 3.25, 3.29, 3.32
   (c) 5.999, 6.002, 6.007, 6.012
   (d) 5.265, 5.269, 5.272, 5.275

30

1.3b   Compare and Order Decimals

Objectives

♦ Compare and order numbers up to 3 decimal places.

Note

This lesson includes some reinforcement for the concepts in the previous lessons. If you have a time constraint, they can be omitted and this lesson combined with the next (or previous) lesson.

Discussion	Text p. 21, Task 4
As with any other set of numbers to compare, we compare 3-place decimal numbers by comparing the digits in the highest place value first. Point out that a 0 was added to 42.45 to have digits in all the places that 42.326 has digits. This makes it easier to line them up to compare. Have students explain their answers. In Task 4(a) we compare the digits in the tenths place to find which is greater, and in 4(b) we compare the digits in the thousandths place. Write a few more pairs of numbers on the board for students to compare. Ask them to rewrite them one on top of the other, aligning the digits in the same places without use of a place-value chart.	4. (a) 42.54 (b) 63.182 42.6, 4.26 4.26 < 42.6 17.39, 17.9 17.39 < 17.9
Assessment	**Text p. 21, Tasks 5-6**

If necessary students can rewrite the numbers vertically, aligning the digits.	5. (a) 3.02, 0.32, 0.302, 0.032
	    (b) 2.628, 2.189, 2.139, 2.045
	
	 6. (a) 0.538, 0.83, 3.58, 5.8
	    (b) 9.047, 9.067, 9.074, 9.076

Reinforcement	
Provide place-value discs (1's, 0.1's, 0.01's, and 0.001's) in a bag. Have students draw about 10 of them and arrange them on a chart. Then have them write a decimal number, a mixed fraction with 1000 in the denominator, the sum of the value of each digit, and the sum of fractions with 10, 100, or 1000 in the denominators. For more challenge have students draw a larger handful so that they are likely to get more than 10 of any type of disc and have to rename some. Then ask them to write the number that is 0.01 more, 0.01 less, 0.001 more, and 0.001 less than the given number. For a bit more challenge have them write the number that is 0.03 more, 0.03 less, 0.003 more, and 0.003 less than their number.	4 ones, 3 tenths, 2 hundredths, 2 thousandths 4.322 $4\frac{322}{1000}$ $4 + 0.3 + 0.02 + 0.002$ $4 + \frac{3}{10} + \frac{2}{100} + \frac{2}{1000}$

Enrichment - Mental Math	
Write the problems at the right and have students use mental math strategies to solve them. For example, in 5.555 + 0.005, since 55 and 5 is 60, then 0.055 and 0.005 is 0.06, so 5.5<u>55</u> + 0.00<u>5</u> = 5.5<u>6</u>.	5.5<u>55</u> + 0.00<u>5</u> 5.5<u>6</u> 5.5<u>55</u> + 0.0<u>6</u> 5.<u>615</u> 5.5<u>55</u> + 0.<u>7</u> 6.<u>255</u> 5.5<u>55</u> − 0.00<u>8</u> 5.5<u>47</u> 5.5<u>55</u> − 0.0<u>9</u> 5.<u>465</u> 5.5<u>55</u> − 0.<u>5</u> 5.<u>055</u>
Enrichment - Mental Math Practice	Mental Math 7-8
Practice	WB Exercise 12, p. 31

Group Game

<u>Material</u>: Place-value discs in a bag (10's, 1's, 0.1's, 0.01's, and 0.001's).

<u>Procedure</u>: Each player draws 10 discs without looking in the bag and writes down the number formed. The player with the greatest number gets a point. The winner is the one who gets 10 points (or some other target number) first.

Exercise 12

1. (a) 4.7   (b) 9.1
   (c) 1.924 (d) 5

2. (a) 624.8
   (b) 5.73
   (c) 1.1

3. (a) >   (b) <
   (c) =   (d) >
   (e) >   (f) <

4. 2.128, 2.18, 2.218, 2.8

5. 6.952, 6.3, 6.295, 6.03

31

1.3c   Decimals to Fractions

Objectives

♦ Express a 3-place decimal as a fraction in simplest form.
♦ Compare and order a mixture of decimals and fractions.

Note

Students will be converting 3-place decimals to fractions by first converting to a fraction with 1000 as the denominator and then simplifying. They will not, however, convert fractions to decimals where the result is a 3-place decimal or larger. All fraction to decimal conversions will result in 1-place or 2-place decimals at this level and so will only involve fractions with denominators that are factors of 100.

Convert 3-place decimals less than 1	Text p. 22, Tasks 7-8

Task 7: Point out that 3-place decimals can be written as thousandths by putting everything after the decimal point over 1000. The 5 is in the hundredths place, but 5 hundredths is the same as 50 thousandths. After students have found the answer, ask them to explain their solution. Point out that although we can realize that both 52 and 1000 can be divided by 4, their greatest common factor, it is easy to simplify in steps. Both 52 and 1000 are even numbers, so we can start by dividing both by 2, which is easy to do mentally. 26 and 500 are also both even, so we can divide both by 2 again. 13 is not divisible by any whole number, so we know that $\frac{13}{250}$ is the simplest form.
 	
Task 8: Have students find the answer. After they are finished ask them if they noticed what factors they ended up dividing the numerators and denominators by in order to simplify them. If they did the simplification in steps using the smallest possible factors, they would only have to use 2 or 5. Any other factors they might have used would have to be factors of 1000. Have students list the factors for 1000. Disregarding 1, all 3-place decimal numbers expressed as a fraction in simplest form will end up with one of these in the denominator. They are all multiples of 5 or 2, so to simplify the fraction with 1000 in the denominator we only need to divide in steps by 2 or 5.
 	
Ask students to express 0.125 as a fraction. The easiest way to do this mentally is divide by 5 continuously. Tell them that the fraction equivalent of 0.125 is useful to memorize, since knowing that 125 x 8 = 1000 will be useful later for mental math.
	7. $\frac{13}{250}$ $\frac{52}{1000} = \frac{26}{500} = \frac{13}{250}$
	
	8. (a) $\frac{1}{2}$
	
	   (b) $\frac{2}{25}$
	
	   (c) $\frac{3}{125}$
	
	   (d) $\frac{69}{200}$
	
	Factors of 1000:
	1, 2, 4, 5, 8, 10, 20, 25, 40, 50, 100, 125, 200, 250, 500, and 1000
	
	$0.125 = \frac{125}{1000} = \frac{25}{200} = \frac{5}{40} = \frac{1}{8}$

Convert 3-place decimals greater than 1	**Text p. 22, Task 9**
Task 9: We only need to convert the decimal part of the number. The resulting fraction will be a mixed number. The fraction can be simplified easily by dividing by 5. Note that although 45 is divisible by 3 and 9, 1000 is not, so we don't even try 3 or 9.	9. $2\frac{9}{200}$

Assessment	Text p. 22, Task 10
	10. (a) $2\frac{3}{5}$ (b) $6\frac{1}{20}$ (c) $3\frac{1}{500}$ (d) $2\frac{51}{125}$
Compare and order decimals and fractions	**Text p. 22, Task 11**
In order to compare decimals and fractions we may need to convert one to the other. Ask students whether they think it would be easier to convert the decimals to fractions or the fractions to decimals. Since they already know what $\frac{4}{5}$ and $\frac{3}{4}$ are as decimals, or should, and it is easy to convert $\frac{7}{25}$ to a decimal, it makes sense to convert the fractions to decimals in order to compare these numbers.	11. (a) 0.6, 0.652, $\frac{4}{5}$, 2 ($\frac{4}{5}$ = 0.8) (b) $\frac{7}{25}$, 0.35, $1\frac{3}{4}$, 7.231 ($\frac{7}{25}$ = 0.28, $1\frac{3}{4}$ = 1.75)
Write the list of fractions and decimals at the right on the board and ask students to put them in increasing order. Discuss any strategies students used. They can all be changed to decimal numbers: 3.2, 2.309, 30.29, 2.39 and then put in order. Here, though, it is not necessary to convert all of them. We can first look at the whole number parts and order them by the whole number. Then only 2.309 and $2\frac{39}{100}$ need to be compared.	$3\frac{1}{5}$, 2.309, 30.29, $2\frac{39}{100}$ In order: 2.309, $2\frac{39}{100}$, $3\frac{1}{5}$, 30.29
Practice	WB Exercise 13, pp. 32-33

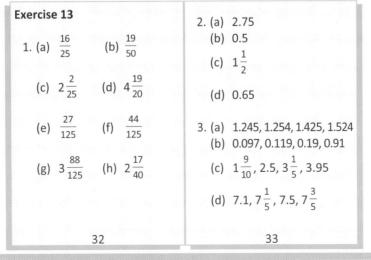

Exercise 13

1. (a) $\frac{16}{25}$ (b) $\frac{19}{50}$

   (c) $2\frac{2}{25}$ (d) $4\frac{19}{20}$

   (e) $\frac{27}{125}$ (f) $\frac{44}{125}$

   (g) $3\frac{88}{125}$ (h) $2\frac{17}{40}$

2. (a) 2.75
   (b) 0.5
   (c) $1\frac{1}{2}$
   (d) 0.65

3. (a) 1.245, 1.254, 1.425, 1.524
   (b) 0.097, 0.119, 0.19, 0.91
   (c) $1\frac{9}{10}$, 2.5, $3\frac{1}{5}$, 3.95
   (d) 7.1, $7\frac{1}{5}$, 7.5, $7\frac{3}{5}$

32

33

Objectives

♦ Practice.

Note

You can have students do the practices independently, in groups, or as a class. You can use these problems to re-teach any concepts that students still need help with.

Practice	Text p. 23, Practice 1A
	1. (a) 0.6 (b) 6 (c) 0.06 (d) 0.006
	2. (a) 4
	(b) 7
	(c) 0.08 or $\frac{8}{100}$
	(d) 0.004 or $\frac{4}{1000}$
	3. (a) 5.509
	(b) 2.819
	(c) 13.52
	4. (a) 0.72 (b) 3.78
	(c) 5.8 (d) 8.04
	5. (a) $\frac{2}{25}$ (b) $\frac{7}{50}$ (c) $\frac{29}{200}$ (d) $\frac{51}{125}$
	(e) $3\frac{3}{5}$ (f) $1\frac{3}{25}$ (g) $4\frac{253}{500}$ (h) $2\frac{3}{500}$
	6. (a) 0.9 (b) 0.03 (c) 0.039 (d) 0.105
	(e) 1.7 (f) 2.18 (g) 3.007 (h) 0.999
	7. (a) 0.07 (b) 0.2
	(c) 2
	(d) 0.005
	(e) $\frac{2}{1000}$ (f) $\frac{4}{1000}$

Practice	Text p. 24, Practice 1B
	1. (a) 0.008, 0.009, 0.08, 0.09 (b) 3.025, 3.205, 3.25, 3.502 (c) 4.386, 4.638, 4.683, 4.9 (d) 9.392, 9.923, 9.932, 10 2. (a) 0.5 (b) 0.75 (c) 0.2 (d) 3.8 (e) 6.25 (f) 4.6 3. (a) = (b) > (c) < (d) = (e) > (f) > 4. (a) 1.703 (b) 0.085 (c) 5.069 (d) 10.052 5. (a) 0.248 (b) 0.792 (c) 3.78 (d) 10.504 (e) 7.009 (f) 9.803

1.4 Rounding Off

Objectives

♦ Round decimal numbers to the nearest whole number.
♦ Round decimal numbers to the nearest tenth.

Material

♦ Meter sticks

Prerequisites

Students should be familiar with rounding whole numbers to the nearest 10 or 100.

Notes

In *Primary Mathematics* 4A students learned to round larger numbers to the nearest ten or hundred. In this part they will learn to round decimal numbers to the nearest whole number or tenth.

Being able to round numbers to a certain place can be used to approximate the answers to addition, subtraction, multiplication, or division problems. Quick approximations are very useful with decimal numbers since it is easy to make a mistake in placing the decimal correctly in the answer. For example, the process for multiplying 42.4 x 0.7 is the same as for 424 x 7, but the answer is 29.68 instead of 2968. We can check the placement of the decimal by rounding the factors to get an approximate answer: 40 x 1 = 40, so the answer will be in tens, not 2.968 or 296.8. Rounding will also be used in division for decimal quotients when the answer contains more decimal places than is needed or is a non-terminating decimal. For example, 46 ÷ 7 = 6.6 to the nearest tenth. In later levels students will also be using estimates of irrational numbers, such as 3.14 for π, and will be rounding answers for area and circumference of a circle.

By convention if a number is exactly halfway between the place the number is being rounded to, it is rounded to the higher number. For example, 465 rounded to the nearest ten is 470.

We can follow the same process in rounding a decimal number as was used in rounding a whole number. To round a number to a specified place, we look at the digit in the next lower place. If it is 5 or greater than 5, we round up. If it is smaller than 5, we round down.

Round 24.25 to the nearest whole number:

24.25 → 24   2 in the tenths place; the number is closer to 24 than 25.

Round 24.25 to the nearest tenth:

24.25 → 24.3   5 in the hundredths place; round up.

Blank Page

1.4a   Round to a Whole Number

Objectives

♦ Round a decimal number to the nearest whole number.

Vocabulary

♦ Rounding
♦ Approximate

Note

Students have already used number lines to understand the process of rounding. Number lines are used in this and the next lesson so that they can understand the process concretely. Even locating a number on a number line requires approximation, particularly if the location is between tick marks, so you may want to include the reinforcement activity that requires students to locate numbers on a number line.

Discussion	Text pp. 25-26, Tasks 1-3
Page 25: Be sure students understand that rounding a number to a particular whole number means to find the whole number that the decimal is closest to. We are finding an approximate value. When we talk about the height of a hill, for example, it hardly matters that it is 3 tenths of a meter more than 164 meters, so we are more likely to say that the hill is about 164 meters high, or even about 160 meters high or about 200 meters high.	1. 37 kg 2. 6 m 3. 25
You can discuss other instances where we might round off a decimal number, such as when estimating the cost of something. Often sellers will mark something as just under a dollar, e.g. $4.99, hoping buyers notice the $4 and won't pay attention to the fact that the price is really almost $5.	
Tasks 1-3: As you discuss rounding these numbers, as well as the one on p. 25, by using their positions on the number lines to see what whole number the decimal is closest to, list each of them, underline the number being rounded to, and write the rounded number. For Task 3, point out that when a number is exactly between two whole numbers, we are simply going to follow an arbitrary rule to round to the higher whole number. This way the 0.5 after the whole number in a number such as 24.5 means to round to 25 and we don't have to care if there are any more digits after the 5, such as in 24.51. Any digit other than 0 after the 5 would make the number closer to 25.	16$\underline{4}$.3 → 164 3$\underline{7}$.4 → 37 $\underline{5}$.78 → 6 2$\underline{4}$.5 → 25
After you are finished with Task 3, have students look at the list you made. Point out that to round a number to a given place, we can look at the number to the right of that place. If it is 5 or more, we round up to the next whole number by adding 1 to the underlined number, the place value we are rounding to, and dropping the remaining digits. If it is less than 5, we round down by simply dropping the rest of the digits.	

Assessment	Text pp. 27, Task 4
	4. (a) 4 (b) 14 (c) 30 (d) 5 (e) 16 (f) 19

Reinforcement	
Draw a number line and mark whole numbers at evenly spaced intervals. Ask students to locate various 1-place and 2-place decimals on it. To do so, they need to estimate how close they are to the whole numbers. For example, on the number line shown at the right, 14.72 will be closer to 15 than 14, and more than halfway from 14. Some students might notice that 0.72 is close to 0.75, which is three quarters, and be able to put it at approximately three quarters from 14 to 15.	14.72? 14 15 16 17

Enrichment	
Ask students for the set of 1-place decimals (tenths) that can be rounded to a given whole number, such as 20.	Tenths that can be rounded to 20: 19.5, 19.6, 19.7, 19.8, 19.9, 20, 20.1, 20.2, 20.3, 20.4
Ask students to round 0.2 m to the nearest meter.	0.2 m rounded to the nearest meter is 0 m.

Practice	
	WB Exercise 14, pp. 34-35

Exercise 14

1. (a) 74

  (b) 10

  (c) 19

  (d) 33

2. (a) 47 lb
  (b) 3 m
  (c) 1 ℓ
  (d) 29 km

3. (a) $3   (b) $11

4. (a) 2 ℓ   (b) 2 ℓ

5. (a) 40   (b) 46
  (c) 6   (d) 6
  (e) 102   (f) 300

34   35

1.4b   Round to a Tenth

Objectives

♦ Round a decimal number to the nearest tenth.

Note

Number lines are again used to help students visualize the process of rounding numbers to one decimal place, or tenths. Provide practice as needed. It is not sufficient for students to simply adapt the procedure taught in the previous lesson of underlining the number we are rounding to and checking the next number to determine whether to round up or down. It is important for them to be able to visualize what is occurring.

Discussion	Text p. 27, Tasks 5-6

As you discuss rounding these numbers using their positions on the number lines, list each of them, underline the number we are rounding to, and write the rounded number.
 	
Task 5: 3.18 m is first rounded to the nearest meter, and then to the nearest tenth of a meter.
 	
Task 6: Point out again that if the number is exactly halfway between two tenths, as in 4.35, we round to the next tenth, 4.4.
 	
After you are finished with these tasks, use the list you made to discuss how to round the numbers without a number line.
	5. (a) 3 m
	   (b) 3.2 m
	
	6. (a) 4.3
	   (b) 4.3
	   (c) 4.4

	3.18 → 3 3.18 → 3.2 4.26 → 4.3 4.32 → 4.3 4.35 → 4.4
Ask students to round 394.65 to the nearest hundred, ten, one, and tenth, without a number line.	394.65 → 400 394.65 → 390 394.65 → 395 394.65 → 394.7
Ask students to round some 3-place decimals to the nearest whole or tenth. They should realize that the thousandths place has no impact or bearing on the process. In rounding to a tenth it is only the digit in the hundredths place that determines which tenth the number should be rounded to.	Round 6.349 to 1 decimal place 6.349 → 6.3

Assessment	Text pp. 27, Task 7
	7. (a) 0.9 (b) 2.5 (c) 7.1 (d) 11.0 (e) 18.0 (f) 24.6
Reinforcment	
You can again have students locate numbers on a number line marked only in tenths. Use a meter stick. The centimeters are hundredths, each 10 centimeters is a tenth. Write a decimal length of less than 1 m. Flip the meter vertically and have students estimate the length on the meter stick, and then check the reverse side to see how close they came.	1.42? ↓ ◄──┼────────┼────────┼────────┼──► 1.4 1.5 1.6 1.7
Enrichment	
Ask students for the set of 2-place decimals that can be rounded to a given 1-place decimal, such as 6.5. Ask students to round 0.03 to the nearest tenth or 1 decimal place.	Hundredths that can be rounded to 6.5: 6.45, 6.46, 6.47, 6.48, 6.49, 6.5, 6.51, 6,52, 6.53, 6.54 0.03 rounded to 1 decimal place is 0.
Practice	WB Exercise 15, p. 36

Exercise 15

1. (a) 4.7
   (b) 8.1

2. (a) 1.5 ℓ
   (b) 20.3 kg
   (c) 9.1 m

3. A: 34.9 kg
   B: 41.7 kg
   C: 39.8 kg

36

Objectives

◆ Review.

Note

There are various ways you can use the reviews. You can do the review in the textbook as a lesson, or do some of the problems for a lesson and assign other problems later for more continuous review if there is extra time in later lessons. Questions 1-5, 8, and 14-16 in Review A review the just-completed unit. The rest of the questions review earlier material in *Primary Mathematics* 4A. Use the review to assess students' understanding and to determine if re-teaching is needed for any topics. Any solutions shown here are simply suggested approaches; students may solve a problem differently.

Review	Text pp. 28-30, Review A
	1. (a) 0.4 (b) 0.02 (c) 3 (d) 100 2. (a) 3.3, 3.03, 0.3, 0.03 (b) 63.5, 6.4, 6.35, 5.63 (c) 0.305, 0.29, 0.05, 0.009 3. (a) 30.06 (b) 16.82 (c) 24.02 (d) 73.2 4. (a) 3 (b) 1 (c) 13 (d) 10 (e) 4 (f) 5 (g) 10 (h) 20 5. (a) 0.8 (b) 0.1 (c) 2.7 (d) 8.1 (e) 10.9 (f) 19.1 (g) 20.6 (h) 10.1 6. (a) 590 (b) 2830 (c) 12,100 7. (a) 5700 (b) 13,800 (c) 45,100 8. (a) 4.05 (b) 4.15

9. (a) 6097
   (b) 364 r2

10. (a) 5700   (b) 10
    (c) 90,000   (d) 320

11. (a) $\frac{2}{3}$ (b) $\frac{1}{2}$

12. (a) $\frac{8}{9} + \frac{8}{9} = \frac{9}{9} + \frac{7}{9} = \mathbf{1\frac{7}{9}}$

    (b) $\frac{2}{3} + \frac{4}{9} = \frac{6}{9} + \frac{4}{9} = \frac{9}{9} + \frac{1}{9} = \mathbf{1\frac{1}{9}}$

    (c) $\frac{5}{8} + \frac{3}{4} = \frac{5}{8} + \frac{6}{8} = \mathbf{1\frac{3}{8}}$

    (d) $3\frac{3}{10}$ (e) $4\frac{7}{10}$ (f) $5\frac{1}{4}$

13. (a) $4 \times \frac{1}{4} = \frac{4}{4} = \mathbf{1}$

    (b) $\frac{1}{5} \times 5 = \frac{1}{5}$ of $5 = \mathbf{1}$

    (c) $7 \times \frac{1}{2} = \frac{7}{2} = \mathbf{3\frac{1}{2}}$

    (d) $5 \times \frac{2}{3} = \frac{10}{3} = \mathbf{3\frac{1}{3}}$

    (e) $\frac{3}{10} \times 8 = 3 \times \frac{8}{10} = 3 \times \frac{4}{5} = \frac{12}{5} = \mathbf{2\frac{2}{5}}$

    (f) $\frac{4}{9}$ of $6 = 4 \times \frac{6}{9} = 4 \times \frac{2}{3} = \frac{8}{3} = \mathbf{2\frac{2}{3}}$

14. (a) 4.03   (b) 1.6   (c) 10.85   (d) 5.75

15. (a) $\frac{4}{5}$ (b) $1\frac{1}{4}$ (c) $4\frac{9}{20}$ (d) $6\frac{3}{50}$

16. (a) 10.4   11.4   12.7
    (b) 2.4   3.4   4.8

17.

    1 unit = 215
    (a) There are 4 more units of pencils than pens.
        4 units = 215 x 4 = **860**
        There are 860 more pencils than pens.
    (b) 6 units = 215 x 6 = **1290**
        Or: 215 + 215 + 860 = 1290
        There are 1290 pens and pencils.

18. Adding 2 fifths 5 times gives (2 x 5) fifths:
    $\frac{2}{5} \ell \times 5 = \frac{2 \times 5}{5} \ell = \frac{10}{5} \ell = \mathbf{2\ \ell}$

    She made 2 ℓ of juice.

29.

    4 units = $20
    1 unit = $20 ÷ 4 = **$5**
    Or: $\frac{1}{4}$ of $20 = $5

    She had $5 left.

30. $\frac{1}{5}$ of the children cannot swim.

    $\frac{1}{5}$ of 40 = **8**

    8 children cannot swim.

Objectives

♦ Review.

Note

It is not necessary to do all three workbook reviews one after the other. Use the reviews at your discretion. You can have students do one page a day as you go on to the next unit, for example, looking through the answers to see if any re-teaching is needed for any particular students.

Review 1

1. 92,405
2. thousands
3. 46,495
4. (a) 6000
   (b) 42,096
   (c) 90,800
   (d) 27,481
5. 78,502
6. 0.03
7. 24,519
8. 30, 60

37

---

9. $\frac{8}{12}$
10. $\frac{7}{12}$
11. 13
12. 3.4
13. A: 1.21 B: 1.28 C: 1.32
14. 4.5, 5
15. 2 km, 20 m, 253 cm, 2 m 35 cm
16. (a) $12
    (b) Sumin

38

---

17. 3 yd − $\frac{5}{6}$ yd = **2 $\frac{1}{6}$ yd**
18. 30 (See solution below.)
19. $\frac{2}{3}$ m x 15 = $\frac{30}{3}$ m= **10 m**
20. 124°
21. angle d
22. 90° − 37° = 53°
23. $25 (See solution below.)

39

18.

```
      15
┌───────────┐
└───────────┘
      ?
```

1 unit = 15
2 units = **30**

23.

```
           $35
┌──┬──┬──┬──┬──┬──┬──┐
└──┴──┴──┴──┴──┴──┴──┘
  shoes      ?
```

7 units = $35
1 unit = $35 ÷ 7 = $5
5 units = $5 x 5 = **$25**

24.

```
Girls ┌──┐  ┐
Boys  ┌──┬─┐ ├1650
           ┘
```

3 units = 1650
1 unit = 1650 ÷ 3 = 550
2 units = 550 x 2 = **1100**
There were 1100 boys.

25. Amount paid in installments
    = 8 x $95 = $760
    $160 + $760 = **$920**
    She paid $920 altogether.

40

26.

```
              36
┌──┬─┬──┬──┐
└──┴─┴──┴──┘
glasses girls   boys
```

3 units = 36
1 unit = 36 ÷ 3 = 12
Half of one unit = **6**
Or: 1 half-unit = 36 ÷ 6 = 6
6 girls wear glasses.

27. 13 cm + 19 cm = 32 cm
    Perimeter: 32 cm x 2 = 64 cm
    Perimeter of square: 64 cm
    Side: 64 cm ÷ 4 = **16 cm**
    Each side of the square is
    16 cm long.

41

Review 2

1. 98,510
2. 1: 10; 100; 1000; 10,000
3. $\frac{6}{10}$ or 0.6
4. 9
5. (a) 48,230
   (b) 70.54
6. $4
7. $\frac{1100 \text{ g}}{2000 \text{ g}} = \frac{11}{20}$

42

---

8. (a) 5.25

   (b) 16.8

9. (a) $\frac{17}{20}$ (b) $2\frac{2}{5}$

10. 5

11. 495

12. 6.3

13. $35

14. (a) 6.05   (b) 3.7

   (c) 0.61   (d) 6.7

15. (a) >   (b) =

16. 10:15 a.m.

43

---

17. 4 km 360 m − 1 km 250 m

   = **3 km 110 m**

18. 98 + 42 = 140

   $\frac{42}{140} = \frac{3}{10}$

19. $\frac{2}{5}$

20. 2 lb − $\frac{1}{4}$ lb = **$1\frac{3}{4}$ lb**

21. $\frac{3}{8}$ of $24 = **$9**

44

---

22. 35 m + 24 m = 59 m

   Perimeter: 2 x 59 m = 118 m

   Cost of fencing:

   118 x $10 = **$1180**

23. 35 yd^2 ÷ 7 yd = **5 yd**

24. 134°

25. (a) CD // IJ

   (b) GH ⊥ PQ

45

---

26. Width = 1 unit

   Perimeter = 6 units = 48 in.

   Length = 2 units

   = 48 in. ÷ 3 = **16 in.**

   The length of the rectangle

   is 16 in.

27.

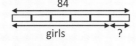

1 unit = boys

6 units = 84

1 unit = 84 ÷ 6 = **14**

There are 14 boys.

46

Review 3

1. P: 89,100

   Q: 89,800

   R: 90,400

2. (a) 6000

   (b) 50,012

   (c) 53,045

3. hundred

4. $15,020

5. $22.50

6. 0.02

7. 5

47

---

8. (a) 21

   (b) 48.02

   (c) 0.06

   (d) 0.2

9. 10.03

10. 40.26, 40.62, 42.06, 42.6

11. 490

12. $\frac{1}{2}$

13. $1\frac{3}{5}$

14. $4

48

---

15. 1 ℓ 500 ml

16. (a) 4650 ml

   (b) 2634 m

   (c) 5107 g

   (d) 184 min

   (e) 4 h 20 min

   (f) 4 kg 7 g

   (g) 5 m 80 cm

   (h) 3 ℓ 20 ml

   (i) 6 lb 12 oz

17. (a) 8 in.

   (b) 32 in.

18. 72 cm^2

19. 2

49

---

20. 150°, 120°

21.

22. 1 kg 680 g / 800 g / } ?

1 kg 680 g − 800 g = 880 g

880 g + 1 kg 680 g = **2 kg 560 g**

The two fruit weigh

2 kg 560 g.

50

---

23. Side of square

   = width of rectangle

   = 20 cm ÷ 4 = 5 cm

   36 cm − 5 cm − 5 cm = 26 cm

   26 cm ÷ 2 = **13 cm**

   The length of rectangle

   ABCD is 13 cm.

24. She cut off $\frac{4}{5}$ m.

   6 m − $\frac{4}{5}$ m = **$5\frac{1}{5}$ m**

   She has $5\frac{1}{5}$ m of thread

   left.

51

---

2   The Four Operations of Decimals

Objectives

- Add or subtract decimals to two places.
- Multiply or divide 1-place and 2-place decimals by a 1-digit whole number.
- Use mental math strategies when appropriate.
- Estimate the sum, difference, product, or quotient in problems involving decimals.
- Round the quotient to one decimal place.
- Solve word problems involving the four operations on decimals.

Suggested number of days: 26

		TB: Textbook WB: Workbook	Objectives	Material	Appendix
2.1	**Addition and Subtraction**				
2.1a	Add Decimals	TB: pp. 31-33 WB: pp. 52-53	◆ Add tenths or hundredths. ◆ Add 1-place decimals.	◆ Place-value discs ◆ Number cubes	◆ Fraction squares (pp. a10-a11) ◆ Mental Math 9
2.1b	Add 2-Place Decimals	TB: pp. 34-35 WB: pp. 54-55	◆ Add 2-place decimals.	◆ Place-value discs	◆ Mental Math 10 ◆ Mental Math 11
2.1c	Estimate Sums	TB: p. 35 WB: p. 56	◆ Estimate the sum for decimals.		
2.1d	Subtract Tenths	TB: p. 36 WB: p. 57	◆ Subtract tenths.	◆ Place-value discs ◆ Number cubes	◆ Mental Math 12
2.1e	Subtract Hundredths	TB: pp. 37-38 WB: pp. 58-59	◆ Subtract hundredths.	◆ Place-value discs	◆ Mental Math 13 ◆ Mental Math 14
2.1f	Subtract Decimals	TB: p. 38 WB: p. 60	◆ Subtract 1-place decimals.	◆ Place-value discs	◆ Mental Math 15
2.1g	Subtract 2-Place Decimals	TB: p. 39 WB: pp. 61-62	◆ Subtract 2-place decimals.	◆ Place-value discs	
2.1h	Estimate Differences	TB: p. 40 WB: p. 63	◆ Estimate differences for decimals. ◆ Use mental math strategies to add or subtract decimals close to a whole.		◆ Fraction squares (pp. a10-a11) ◆ Mental Math 16
2.1i	Word Problems	TB: pp. 41-42 WB: pp. 64-66	◆ Solve word problems involving addition or subtraction of decimals.		
2.1j	Practice	TB: pp. 43-44	◆ Practice.		

		TB: Textbook WB: Workbook	Objectives	Material	Appendix
2.2	**Multiplication**				
2.2a	Multiply Tenths and Hundredths	TB: pp. 45-47 WB: pp. 67-68	♦ Multiply tenths. ♦ Multiply hundredths.	♦ Place-value discs	♦ Mental Math 17
2.2b	Multiply Decimals	WB: p. 69	♦ Multiply 1-place decimals by a whole number.	♦ Place-value discs	
2.2c	Estimate Products	TB: pp. 48-49 WB: pp. 70-71	♦ Multiply 2-place decimals by a whole number.	♦ Place-value discs	♦ Mental Math 18
2.2d	Word Problems	TB: pp. 50-51 WB: pp. 72-74	♦ Solve word problems involving multiplication of decimals.		
2.2e	Practice	TB: p. 52	♦ Practice.		
2.3	**Division**				
2.3a	Divide Decimals	TB: pp. 54-55 WB: pp. 75-76	♦ Divide decimals using division facts.	♦ Place-value discs	♦ Mental Math 19
2.3b	Divide Hundredths	TB: p. 56 WB: pp. 77-78	♦ Divide 2-place decimals by a 1-digit whole number where the quotient is less than 1.	♦ Place-value discs	
2.3c	Divide 2-Place Decimals	TB: p. 57 WB: pp. 79-80	♦ Divide 2-place decimals by a 1-digit whole number.	♦ Place-value discs	
2.3d	Append 0's to Divide	TB: p. 58 WB: pp. 81-82	♦ Divide decimals by adding decimal places when there is a remainder.	♦ Place-value discs	
2.3e	Estimate and Round Quotients	TB: p. 58 WB: p. 83	♦ Estimate the quotient for division of decimals. ♦ Round the quotient to one decimal place.		
2.3f	Word Problems	TB: pp. 60-61 WB: pp. 84-86	♦ Solve word problems involving division of decimals.		
2.3g	Practice	TB: p. 62	♦ Practice.		
2.3h	Practice	TB: p. 63	♦ Practice.		♦ Mental Math 20
2.3i	Practice	TB: p. 64	♦ Practice.		
2.3j	Review	TB: pp. 65-67	♦ Review.		

2.1 Addition and Subtraction

Objectives

♦ Add or subtract decimals to two places.
♦ Use mental math strategies when appropriate.
♦ Estimate the sum or difference of decimals.
♦ Solve word problems involving addition and subtraction of decimals.

Material

♦ Place-value discs for 0.001, 0.01, 0.1, 1, 10, 100
♦ Fraction squares for whole, tenths, hundredths
♦ Number cubes (0.4, 0.5, 0.6, 0.7, 0,8, 0.9)

Prerequisites

Students should be able to add and subtract whole numbers of at least 4 digits using the standard algorithms and be comfortable with place-value concepts. They should also be familiar with mental math strategies used with 2-digit whole numbers and mental math strategies for adding or subtracting numbers close to 100.

Notes

Students learned and practiced the standard algorithms for addition and subtraction of whole numbers in *Primary Mathematics* 2A, 3A, and 4A. The standard algorithms for addition and subtraction of decimals are the same as those for whole numbers. Numbers are aligned vertically by place and addition or subtraction is done starting from the smallest place, that is, from right to left. In the process, the value at any place may have to be renamed.

The standard algorithms with decimals will be introduced using simple addition or subtraction problems where one or both numbers have only one non-zero digit. Simpler problems give students a bridge to use the algorithms correctly for problems that cannot be easily solved mentally, such as numbers with more than two non-zero digits. In the exercises you can allow capable students to use mental math strategies when possible. However, during the lesson, encourage even more capable students to write some problems vertically in order to practice aligning the digits and the decimal point correctly.

When discussing the steps in using any of the standard algorithms use place-value names. For example, in using the standard algorithm for 9.79 + 4.86, we first add 9 *hundredths* and 6 *hundredths* to get 15 *hundredths*, which is the same as 1 tenth 5 hundredths. Do not say, "we first add 9 and 6."

Students learned to round whole numbers to estimate answers to problems in *Primary Mathematics* 4A. In Unit 1 of *Primary Mathematics* 4B they learned to round decimal numbers. In this part students will learn to estimate answers to problems involving decimal numbers. Encourage them to make a habit of estimating answers. Estimation helps reduce errors that may result from putting the decimal point in the wrong place and can help them to determine whether an answer is reasonable. Estimation is particularly useful in multiplication and division where there are more places for potential errors, and will be useful later when students multiply or divide a decimal by a decimal.

When using estimation to determine whether an answer is reasonable, there is no hard and fast rule for what place to round the numbers to, and to impose rules defeats the purpose of estimation: to give an *approximate* answer. There is no *exact* answer for an estimate. Students should round to numbers that allow them to find the estimated answer quickly. This could be to the same place value for some students, such as 25.48 + 7.64 = 25 + 8, if they find it easy to mentally add 25 and 8. But they could also round the numbers to 30 + 8, and that is sufficient to check the reasonableness of the actual answer. Any estimates provided in this guide as answers are just possible estimated answers; students might have a more or less precise answer depending on how they rounded the numbers.

In *Primary Mathematics* 3A students learned mental addition and subtraction of numbers close to 100, or close to a ten, and applied those strategies to money. For example:

$4.55 + $1.99 = $6.54
    Add $2, and then take off 1¢.

$4.51 − $1.97 = $2.54
    Subtract $2, and then add 3¢.

These strategies will be extended to decimal numbers in this part.

At this level students will only encounter addition and subtraction problems for 2-place decimals. The algorithm for adding and subtracting numbers to any number of decimal places, or whole number places, is the same.

Students will be solving two-step word problems involving adding or subtracting decimals. In *Primary Mathematics* 3A they learned how to use the part-whole and comparison models to solve word problems involving addition and subtraction. The drawing in Task 35 on p. 41 is an example of a part-whole model, and the one in Task 36 on the same page is an example of a comparison model. Modeling is used, but not re-taught here. If necessary, review the model method as it is taught in the *Primary Mathematics* 3A textbook.

The purpose of the model method is to provide a way for students to translate the words in the problem into a visual picture, from which they can see what equations they must use. Understanding these models will be useful to solve more complex word problems in any of the supplementary books and in later levels of *Primary Mathematics.* Use the models during lessons, but use your discretion in requiring them for independent work; do not force more capable students to draw models every time for routine problems if they can easily solve the problem without one. Model drawing is a means to an end, not a means in and of itself. Try to provide more capable students with more complex word problems from some of the supplementary books so they can see how useful the models are for such problems. For other students, if they do have difficulties with the word problems, have them draw the models more frequently, particularly in correcting problems they get wrong for other reasons than computation error.

Do not get caught up in the process of drawing the models and require students to follow a set of arbitrary steps in creating them in such a way that the process becomes more important than the end result (the answer to the problem). Do not establish a set of steps to be applied to all problems. Any such set of steps that works with some problems may not be the best way to approach a different problem. Forcing a specific set of steps will short-change logical thinking and the development of problem-solving skills in the long run, and diminishes exploration of alternate approaches.

The purpose of model drawing is for students to be able to determine an approach to solve the problem. Some students can determine an appropriate approach with a simple model or a partial model. Do not hinder this use of models by requiring a specific type of model or method of labeling. Notice that the models in the textbook are not always labeled with what or who they represent, or always with a question mark for what is supposed to be found. Students should not be graded on whether they drew a model, how neat the model is, or whether it was labeled precisely.

Have students give the answer to word problems in a statement. This helps them to determine whether they actually answered the question posed in the problem. Do not have them write the statement with a blank for the answer before they even work on the problem, though. Some students can write the statement, and still just fill in the blank with their solution without re-reading it.

Solutions using a model provided in this guide are just suggested solutions to help you with an idea of where to "lead" your students if they are having trouble — other solutions are possible. Encourage students to discuss alternate approaches.

2.1a   Add Decimals

Objectives

♦ Add tenths or hundredths.
♦ Add 1-place decimals.

Note

Most students will not need an extensive concrete introduction to adding decimals. Place-value discs should be sufficient. If some do need to visualize the process more concretely, you can illustrate the process with fraction squares or base-10 blocks where the flats are ones.

If a student has messy handwriting, he or she can use graph paper or lined paper turned sideways to help align the digits.

Add tenths or hundredths	Text p. 31, p. 32, Tasks 1-3
Page 31: This page illustrates adding and subtracting tenths using measurement as the concrete introduction. In this example there is no renaming. Tenths are added to tenths.	$0.7 + 0.2 = \mathbf{0.9}$ They drank **0.9** liter of milk together. $0.7 - 0.2 = \mathbf{0.5}$ David drank **0.5** liter more than John.
Task 1: This task moves to a more pictorial representation of addition of tenths to tenths and extends the concept to adding hundredths to hundredths. You can introduce this task more concretely by providing students with actual place-value discs. Tenths are added to tenths, and hundredths to hundredths, in the same way that ones are added to ones.	1. (a) 0.7 (b) 0.07 2. 1.3 3. 0.13
Tasks 2-3: These tasks illustrate addition of tenths or hundredths when the answer is more than 10 tenths or 10 hundredths. Students should have no difficulty with the concept of renaming 10 tenths as 1 whole, or 10 hundredths as 1 tenth. You can illustrate the problems with actual place-value discs if needed. Write the problems horizontally and have students rewrite them vertically. Point out that we can write the numbers vertically to keep track of the place value of each digit. When we write the numbers this way, it is important to align the decimal point. That way, tenths are aligned with tenths and hundredths with hundredths.	
Assessment	**Text p. 33, Task 4**
Most students can probably answer these mentally.	4. (a) 0.8 (b) 1.3 (c) 1.2 (d) 0.06 (e) 0.1 (f) 0.17

Add 1-place decimals	Text p. 33, Tasks 5-6
Task 5: This task shows that you can split a number into the whole and fractional part and add tenths in the same manner as in the previous lesson and then add the result to the whole number part. Discuss mental math strategies. We can also add 0.4 to 6.9 by making the next whole.	5. 7.3 6. 5.4 $6.9 + 0.4 = 7 + 0.3 = 7.3$ $\quad\quad\diagup\diagdown$ $\quad 0.1\ \ 0.3$
Task 6: This task shows the standard algorithm for addition. Copy the problem on the board and go through the steps for solving it, using place-value discs if needed. Be sure to use place-value names for each digit when discussing the steps. We first add the digits in the smallest place value, which in this case is tenths. Since there are 14 tenths, we rename them as 1 one and 4 tenths, and write the 1 one above the ones column to remember to add it to the rest of the ones. Discuss ways to solve this problem mentally, such as adding the ones first and then the tenths. We are adding 36 tenths and 18 tenths. We can add them the same way as adding 36 ones and 18 ones. We then have to remember that the answer is tenths, not ones, and put in the decimal point accordingly.	$3.6 + 1.8 = ?$ $36 + 18 = 46 + 8 = 54$ $\quad\quad\quad\quad\diagup\diagdown$ $\quad\quad\quad\ 4\ \ \ 4$ $3.6 + 1.8 = 4.6 + 0.8 = 5.4$ $\quad\quad\quad\quad\diagup\diagdown$ $\quad\quad\quad 0.4\ \ 0.4$
Assesment	**Text p. 33, Task 7**
Allow students to use either the standard algorithm or mental math, depending on what they need to work with the most.	7. (a) 8.5 (b) 3.5 (c) 4 (d) 9.6 (e) 6 (f) 6.7
Enrichment - Mental Math	Mental Math 9
Practice	WB Exercises 16-17, pp. 52-53

Group Game

Material: Place-value discs (10's, 1's, and 0.1's). Number cube with 0.4, 0.5, 0.6, 0.7, 0.8, and 0.9.

Procedure: Each player starts with five 1-discs and five 0.1-discs. They write down 5.5. Players take turns throwing the number cube. They collect 0.1-discs according to the number showing face-up on the number cube, trading in ten 0.1-discs for a 1-disc as needed. You can have the player write an addition equation for each throw. The player who collects a 10-disc first wins.

Exercise 16	Exercise 17

1. (a) 0.8
   (b) 1.2
   (c) 0.6
   (d) 1
   (e) 1.4
 	
2. (a) 0.06
   (b) 0.12
   (c) 0.05
   (d) 0.1
   (e) 0.11
	1. (a) 3.1
	   (b) 5.4
	   (c) 10.5
	   (d) 6.2
	
	2. (a) 5 (b) 8.3
	   (c) 13.7 (d) 16.3

52	53

2.1b   Add 2-Place Decimals

Objectives

◆ Add 2-place decimals.

Note

In this lesson students will add decimal numbers mentally or use the standard algorithm. Allow them to pick the method they want to use. The goal is to get the correct answer, and for some students the standard algorithm can be more effective for some problems. The type and number of problems a student can comfortably solve mentally with assurance of a correct answer can vary between students. However, encourage discussion of and use of mental math strategies.

Discussion	Text pp. 34-35, Tasks 8-9, 11
Write the problems in Task 8, 9, and 11 and discuss them with students. Task 8: This task shows that when we add tenths or hundredths to a 2-place decimal we add the tenths to the tenths and the hundredths to the hundredths. Discuss alternate mental math strategies. For Task 8(b), we could make the next tenth with either 0.42 or 0.09, taking the hundredths from the other number. We could also, in this case, add a tenth and subtract a hundredth, since 0.09 is one hundredth less than a tenth. Task 9: This task shows the standard algorithm for addition. Review the steps, using place-value discs if needed. We first add the digits in the smallest place value, which in this case is hundredths. Since there are 11 hundredths, we rename them as 1 tenth and 1 hundredth and write the 1 tenth above the tenths column to remind us to add it to the rest of the tenths. Discuss ways to solve the problem in Task 9 mentally. We are adding 24 hundredths and 37 hundredths. We can add them the same way we add 24 ones and 37 ones. We then have to remember that the answer is hundredths, not ones, and put in the decimal point accordingly. Task 11: In this task, we are adding the equivalent of 3-digit numbers. Go through the steps in the algorithm, using place-value discs. Remind students that when the numbers have more digits, it can be easier to rewrite the number vertically and add starting with the digits in the smallest place value than trying to solve the problem mentally, particularly if renaming occurs over several places.	8. (a) 1.32 (b) 0.51 9. 0.61 11. 6.47 $$0.42 + 0.09 = 0.5 + 0.01 = 0.51$$ $$0.08 \quad 0.01$$ $$0.42 + 0.09 = 0.41 + 0.1 = 0.51$$ $$0.41 \quad 0.01$$ $$0.42 + 0.09 = 0.42 + 0.1 - 0.01$$ $$= 0.52 - 0.01$$ $$= 0.51$$ $$0.24 + 0.37 = \,?$$ $$24 + 37 = 54 + 7 = 61$$ $$6 \quad 1$$ $$0.24 + 0.37 = 0.54 + 0.07 = 0.61$$ $$0.06 \quad 0.01$$
Write the expression **49.79 + 4.86** on the board and ask students to rewrite it vertically, aligning the digits, and add. Review the steps for addition, which are the same as with whole numbers.	$49.79 + 4.86$ $$\begin{array}{r} {\scriptstyle 1\ \ 1\ \ \ 1} \\ 49.79 \\ +\ \ 4.86 \\ \hline 54.65 \end{array}$$

Repeat with **68.3 + 6.98** and **45.97 + 589**. Point out that in order to have the same number of digits after the decimal, we can add 0's. Note that if there is no fractional part to the number, we need to include a decimal point before adding 0's. Just adding 0's to 589 without a decimal point changes the value of the number, since that will push the digits to a higher place, but writing a decimal point and then adding 0's does not change the value.	68.3 + 6.98 $$\begin{array}{r}{\scriptstyle 1\ \ 1}\\ 68.30\\ +\ \ 6.98\\ \hline 75.28\end{array}$$ 45.97 + 589 $$\begin{array}{r}{\scriptstyle 1\ \ 1}\\ 45.97\\ +589.00\\ \hline 634.97\end{array}$$
Assessment	**Text p. 34, Task 10**
Students can solve these mentally, or rewrite them vertically and use the standard algorithm.	10. (a) 2.63 (b) 0.96 (c) 1.14 (d) 6.02 (e) 0.4 (f) 1.03 (g) 4.28 (h) 1.18 (i) 1.35 (j) 7.49 (k) 3.06 (l) 4
Provide some additional problems with more digits, such as those shown at the right. You can have several students come to the board and compete to see who can find the correct answer first.	7.89 + 68.4 76.29 55.87 + 8.57 64.44 29.07 + 8.8 37.87
Enrichment - Mental Math Practice	Mental Math 10-11
Enrichment	
Give students some numbers to add that go out to more than two decimal places or involve more than 4 digits in the number. They should realize that the standard algorithm can be applied to numbers with digits in any number of places, whether larger than millions or smaller than thousandths.	56.9831 + 43.5 = 100.4831 4,568,192.872 + 4,458.98 = 4,572,651.852
Practice	WB Exercise 18, pp. 54-55

Exercise 18

1. (a) 2.73
   (b) 2.55
   (c) 5.05
   (d) 4.57
   (e) 6.24
   (f) 3.88
   (g) 2.7
   (h) 4.34

2. (a) 0.92 (b) 3.03
   (c) 2.36 (d) 28.28
   (e) 3.62 (f) 9.61
   (g) 17.34 (h) 68.18

54

55

2.1c Estimate Sums

Objectives

♦ Estimate the sum for decimals.

Note

There is no exact rule for how to round a number when performing estimation. If students are incapable of determining what digit to round to under what circumstances, then you may have to impose some arbitrary rules. However, requiring a set of rules to generate only one possible answer to estimation indicates that students lack number sense, so it is worth spending time on different potential estimations to help encourage a better number sense.

Discuss estimation	
Write the expression **$4.80 + $2.37** on the board. Remind students that we can estimate the answer to a problem by rounding each number that we are adding together. How we round depends on the situation. If, for example, we wanted to estimate if we had enough money to buy two items that cost $4.80 and $2.37, we might want to round both numbers up, that is, $5 + $3, rather than the second number to the nearest whole number, $2.	$4.80 + $2.37 = ? $5 + $3 = $8; I need about $8. $5 + $2 = $7; the sum will be close to $7. $4.80 + $2.37 = $7.17
Write the expression **34.26 + 10.82**. Tell students that if we want a quick estimate to see if an exact answer makes sense or has the decimal point in the right place, we could to round to a number with a single non-zero digit. We would round 34.26 + 10.82 to 30 + 10 to give the estimated answer of 40. Sometimes it is helpful to have a closer estimate. Since it is easy to add 2-digit numbers mentally, we could round the numbers in this example to the nearest whole number. This gives a closer estimate to the actual value. Ask students to find the actual sum.	34.26 + 10.82 = ? 30 + 10 = 40; the sum will be about 40. 34 + 11 = 44; the sum will be close to 44. 34.26 + 10.82 = 45.08
Write the equation **245.3 + 39.02 = 63.55** and ask students if the answer makes sense. Obviously it does not, since we are adding to hundreds the answer should be hundreds. We can simply use the estimate 200 + 40 to realize the answer is incorrect. Ask students what error might have led to the incorrect answer. Probably the digits were not aligned correctly.	245.3 + 39.02 = 63.55 ?
Then write **245.3 + 39.02 = 294.32** and ask students if this is correct. We could estimate 250 + 40 = 290 so it could be correct. It is not, however. Ask students to find the correct answer. Estimation in addition or subtraction is a good way to do a quick check to make sure digits are aligned correctly, but will not tell us whether the answer is indeed correct. Even so, we should always make sure their answer *makes sense*, and a quick estimate with whatever is easy to add mentally is a good habit. Even just recognizing that if we are adding to 200 the answer cannot be around 60 is a good use of estimation.	245.3 + 39.02 = 294.32 ? 245.3 + 39.02 = 284.32 ✓

Discussion	Text p. 35, Tasks 12-13
Task 12: Ask students to write down a quick estimate of the sum of both problems, and then solve for the exact answer.	12. (a) 33.12 (b) 7.17 33.12 7.17
Task 13: We could round to the nearest whole number and use mental math to find the estimated value as shown. Or, we could round the numbers to 30 and 10, or even simply add 10 to 34. Rounding both to the nearest whole number gives a more precise estimate, but rounding to one non-zero digit is sufficient to determine if the answer is reasonable with respect to the placement of the decimal point.	13. 45, or 40, or 44
Write the expression **59.3 + 0.67**. Ask students how we should round the numbers to get an estimate. 0.67 is so much smaller than 59 that it is obvious that the answer won't be much more than 59, so we don't really have to do any agonizing over how many digits to round the numbers to in order to get an estimate.	$59.3 + 0.67 = ?$
Assessment	**Text p. 35, Task 14**
Ask students to first write an estimate, and then find an exact answer. Accept any reasonable estimates.	14. (a) 9.76 (b) 6.34 (c) 7.18 (d) 5.92 (e) 9.43 (f) 13.08
Practice	WB Exercise 19, p. 56
It is up to you whether you want to require students to find exact answers for this exercise. A capable student may realize that he can determine what letters to write in which space using primarily estimation, and find exact answers only to distinguish between numbers that are close.	

Exercise 18

1. D 42.9   S 20.51
     Q 44.09   M 90
O 11.36   N 66.9
     M 33.6   U 63
I 27.35   J 88.75
    A 68.05   V 82

VANDA MISS JOAQUIM

56

2.1d   Subtract Tenths

Objectives

♦ Subtract tenths.

Note

If students are competent with mental math and subtraction, you may want to combine this lesson with the next lesson.

Discussion	Text p. 36, Tasks 15-16
Write the problems on the board and discuss them. Use place-value discs to illustrate the process if needed. Students can probably do these problems mentally. Emphasize the renaming process.	15. (a) 0.6 (b) 0.8 (c) 2.8 16. 3.4
Task 15(a): This task shows that we subtract decimals in the same way that we subtract whole numbers. We can subtract 2 tenths from 8 tenths in the same way that we can subtract 2 ones from 8 ones.	
Task 15(b): To subtract tenths from ones we need to rename a one as 10 tenths. So we end up subtracting 2 tenths from 10 tenths.	

Task 15(c): We can think of this problem as subtracting 2 tenths from 30 tenths, and solve in the same way as we would solve 30 – 2, but remember that these are tenths, not ones. Or, we can think of this problem as renaming only one of the tenths, and subtract 2 tenths from 10 tenths, remembering we still have 2 ones.	$3 - 0.2 = 2 + 0.8 = 2.8$
	   / \\
	  2 1.0

Task 16: This task shows the standard algorithm for subtraction. Discuss the steps with students. We do not have enough tenths to subtract 8 tenths, so we rename a one as 10 tenths, giving us 12 tenths. Then we subtract 8 tenths from that.	$4.2 - 0.8 = ?$ $42 - 8 = 2 + 32 = 34$ / \\ 2 40
Discuss ways to solve the problem mentally, such as subtracting 8 tenths from a one and adding the 2 tenths back in, or, since 8 tenths is almost 1, subtracting 1 and adding back in 2 tenths.	$4.2 - 0.8 = 0.2 + 3.2 = 3.4$ / \\ 0.2 4
Point out that if we are only subtracting tenths, we can ignore any hundredths, and just include that in the answer.	$4.2 - 0.8 = 4.2 - 1 + 0.2 = 3.4$
Write the problem **4.26 – 0.8** and have students find the answer. Discuss both the mental math and the standard algorithm strategies.	$4.26 - 0.8 = 3.4 + 0.06 = 3.46$ / \\ 0.06 4.2 $$\begin{array}{r} \overset{3}{\cancel{4}}.\overset{1}{2}6 \\ -\ 0.8 \\ \hline 3.46 \end{array}$$
Provide additional examples as needed. You can use some of the problems in Task 17.	

Assessment	Text p. 36, Task 17
	17. (a) 0.2 (b) 0.2 (c) 0.7 (d) 0.6 (e) 1.3 (f) 3.1 (g) 0.6 (h) 4.1 (i) 4.4 (j) 0.28 (k) 3.55 (l) 4.72
Mental Math Practice	Mental Math 12
Practice	WB Exercise 20, p. 57

Group Game

Material: Place-value discs (1's, and 0.1's). Number cube with 0.4, 0.5, 0.6, 0.7, 0.8, and 0.9.

Procedure: Each player starts with five 1-discs and five 0.1-discs. They write down 5.5. Players take turns throwing the number cube. They take away 0.1-discs according to the number showing face-up on the number cube, trading in a 1-disc for ten 0.1-discs when necessary. You can have the player write a subtraction equation for each throw. The player who gets rid of all his discs first (or does not have enough to subtract from) wins.

Exercise 20

1. (a) 0.6
  (b) 0.9
  (c) 0.3
  (d) 3.9

2. (a) 5.3   (b) 2.6
  (c) 3.16   (d) 2.2

57

Objectives

♦ Subtract hundredths.

Note

This lesson should be easy for students who are competent with mental math. Students who are not can use the standard algorithm. The emphasis should be on paying attention to the place value of the digits being subtracted and on renaming; ones to tenths and tenths to hundredths. For students who have difficulty use place-value discs on a place-value chart.

Discussion	Text pp. 37-38, Tasks 18-19, 21
Write the problems on the board and discuss them. Use place-value discs to illustrate the process.	18. (a) 0.02 (b) 0.04 (c) 0.94
Task 18(a): This task shows that we subtract hundredths from hundredths in the same way as we subtract whole numbers. 8 hundredths − 6 hundredths = 2 hundredths.	19. 0.77 21. 4.12
Task 18(b): To subtract hundredths from tenths, we need to rename a tenth as 10 hundredths.	
Task 18(c): To subtract hundredths from ones, we need to rename a one as 10 tenths, and one of those tenths as 10 hundredths. Point out that 1 is 100 hundredths. We can use mental math strategies for making 100.	$1 - 0.06 = ?$ $100 - 6 = 94$ $1.00 - 0.06 = 0.94$
Task 19: To subtract tenths and hundredths from 1, we can rename the one as 9 tenths and 10 hundredths, and then subtract the tenths from 9 tenths and the hundredths from 10 hundredths. We use the same strategies as we used for making 100.	$1 - 0.23 = ?$ $100 - 23 = 77$ $1.00 - 0.23 = 0.77$
Task 21: We can also use the standard algorithm to subtract. Point out that 4.2 does not have any hundredths, but that we are subtracting hundredths. When we rewrite this problem vertically, we include a 0 for the hundredths digit. After renaming a tenth we have 10 hundredths.	
Point out to students that when rewriting the problem vertically, we have to be careful to align the digits correctly, not like the incorrect example shown at the right. If we have done so, the decimal point is also aligned.	$$\begin{array}{r} 4.2 \\ -\ 0.08 \\ \hline \end{array}$$ (crossed out)
	$4.26 - 0.08 = ?$ $$\begin{array}{r} \overset{1}{4.\overset{1}{2}6} \\ -\ 0.08 \\ \hline 4.18 \end{array}$$
Write the expression **4.26 − 0.08** vertically and discuss its solution using the standard algorithm. This time, after renaming a tenth, we have 16 hundredths. Then discuss mental math strategies. Point out that we can also use the same mental math strategies we would use for 26 − 8.	$26 - 8 = 6 + 12 = 18$ 　　/＼ 　6　20 $4.26 - 0.08 = 4.18$ 　　/ ｜ ＼ 　4 0.06 0.20

Assessment	Text pp. 37-38, Tasks 20, 22

	20. (a) 0.07 (b) 0.47 (c) 3.47
	     (d) 0.06 (e) 0.26 (f) 2.26
	     (g) 0.93 (h) 1.93 (i) 3.91
	     (j) 0.55 (k) 2.55 (l) 3.14
	
	22. (a) 3.23 (b) 3.47 (c) 4.16
	     (d) 4.74 (e) 6.13 (f) 6.41

Mental Math Practice	Mental Math 13-14
Practice	WB Exercise 21, p. 58-59

Exercise 21

1. (a) 0.05
   (b) 0.65
   (c) 0.85
   (d) 0.92

2. (a) 4.38
   (b) 1.48

3. (a) 0.42   (b) 3.24

   (c) 2.78   (d) 6.06

   (e) 2.62   (f) 4.23

   (g) 5.04   (h) 3.91

58

59

2.1f   Subtract Decimals

Objectives

♦ Subtract 1-place decimals.

Note

If students are competent with subtraction and place-value concepts, you may want to combine this lesson with the next lesson.

Illustrate the subtraction algorithm	
Write the expression **5.2 − 2.7** horizontally on the board. Have students rewrite the problem vertically and go through the steps of the subtraction algorithm, using place-value discs to illustrate renaming. Start with the tenths. We cannot subtract 7 tenths from 2 tenths, so we rename one of the ones as 10 tenths, which gives us 12 tenths. 12 tenths − 7 tenths = 5 tenths. We then subtract the 2 ones from the remaining 4 ones to get 2 ones.	5.2 − 2.7 $$\begin{array}{r} \overset{4}{\cancel{5}}.\overset{1}{2} \\ -\ 2\ .\ 7 \\ \hline 2\ .\ 5 \end{array}$$
Discuss ways to solve this problem mentally. 5.2 is the same as 52 tenths. And 2.7 is the same as 27 tenths. We can subtract mentally the same way we subtract 27 from 52, but the answer is tenths.	5.2 − 2.7 = ? $$52 - 27 = 32 - 7 = 23 + 2 = 25$$ $$\underset{2\quad 30}{\wedge}$$ 5.2 − 2.7 = 2.5
Repeat with **45 − 8.9**. Make sure students align the digits correctly. In this case, we can add a decimal point and 0 to 45, since we will need to rename a one as 10 tenths. You can discuss mental strategies if you want, but this problem is more complicated since if we renamed everything as tenths we would be doing 450 tenths − 89 tenths, which is harder to do mentally than 52 tenths − 27 tenths. One possibility is to subtract 90 tenths and add 1 tenth, since 450 − 90 is relatively easy to do mentally. Remind students that they can use mental math strategies when they are comfortable doing so, but sometimes simply rewriting vertically and using the standard algorithm is actually faster than trying to figure out what mental math strategy to use.	45 − 8.9 $$\begin{array}{r} \overset{3}{\cancel{4}}\ \overset{1}{\cancel{5}}.\overset{1}{0} \\ -\quad 8\ .\ 9 \\ \hline 3\ 6\ .\ 1 \end{array}$$
Discussion	**Text p. 38, Task 23**
Copy the problem on the board and discuss the steps in solving it. There are no tenths to subtract 7 from, se we rename a one as 10 tenths, adding a decimal point and a 0 after the 6.	23. 3.3
Discuss mental math strategies. This problem is easy to solve mentally for some students, as long as we keep track of place values and remember that we have 60 tenths.	6 − 2.7 = ? 60 − 27 = 33 6.0 − 2.7 = 3.3

Assessment	Text p. 38, Task 24
	24. (a) 3.6 (b) 3.5 (c) 2.7 (d) 2.5 (e) 2.6 (f) 4.8
Enrichment - Mental Math Practice	Mental Math 15
Practice	WB Exercise 22, p. 60

Exercise 22

1. (a) 2.1   (b) 2.7

   (c) 3.6   (d) 1.6

   (e) 2.2   (f) 1.4

   (g) 4.1   (h) 3.6

60

2.1g Subtract 2-Place Decimals

Objectives

♦ Subtract 2-place decimals.

Note

Although some students may be capable of mental math strategies with 3-digit numbers, there is more chance for error, and with decimals more chance for errors related to place values. You can allow students to use mental math strategies or not, depending on your classroom situation. Using the standard algorithm should be easy and automatic. Students who do struggle with mental math strategies or deciding what strategy to use should always be comfortable with using the standard algorithm.

Discussion	Text p. 39, Tasks 25-26

Write the problems on the board and guide students through the steps.
 	
Task 25: To subtract the tenths, we need to rename a one as 10 tenths. The process is the same as if we were subtracting two 3-digit numbers. However, in copying the problem, it is important to remember to include the decimal point, since we are subtracting ones, tenths, and hundredths, not hundreds, tens, and ones.
 	
Task 26: You may want to have a student come to the board for the problems in Task 26 to explain the steps. When we rewrite the problems vertically, we can add trailing zeros in order to have the same number of digits after the decimal point. This makes it easier to align the digits correctly. For example, in 26(a), there are hundredths in the first number (7.24), but not the second (3.5), so we add a 0 for hundredths (3.50). Emphasize that it is important to remember to include the decimal point when copying the numbers, since they should also align.
	25. 1.74
	
	26. (a) 3.74 (b) 0.31
	      3.74 0.31
	   (c) 3.73 (d) 2.66
	      3.73 2.66

Assessment	**Text p. 39, Task 27**

	27. (a) 0.42 (b) 0.25 (c) 0.88
	    (d) 3.4 (e) 3.49 (f) 3.55
	    (g) 0.44 (h) 2.15 (i) 1.62
	    (j) 1.55 (k) 3.44 (l) 0.95

Enrichment

After students have finished Task 27 and checked answers, you may want to ask them how they solved the problems. Have them explain their solutions if they used mental math strategies.	
For example, 0.85 – 0.43 in 27(a) can easily be done mentally using strategies for 2-digit numbers, and does not even involve renaming. Similarly for 0.64 – 0.39 in 27(b), except that renaming is needed. Note that we cannot do 2.9 – 0.75 in 27(h) as we would 29 – 75 because of the placement of the decimal point. Since students have learned to do a problem such as 1 – 0.56 using strategies for making 100, 6 – 2.56 in 27(k) could be done mentally by splitting 6 into 5 and 1, and remembering to subtract 2 from the 5, or first subtracting the 2 and then 0.56 from one of the ones.	$0.85 - 0.43 = 0.42$ $0.64 - 0.39 = ?$ $64 - 39 = 34 - 9 = 21 + 4 = 25$ $\wedge$ 4 30 $0.64 - 0.39 = 0.25$ $1 - 0.56 = 0.44$ $6 - 2.56 = 4 - 0.56 = 3.44$ $\wedge$ 3 1
Give students some numbers to subtract that go out to more than two decimal places or involve more than 4 digits in the number. Students should realize that the standard algorithm can be applied to numbers with digits in any number of places, whether larger than millions or smaller than thousandths.	$52.03 - 3.5127 = 48.5173$ $4,500,102.322 - 4,458.98 = 4,495,643.342$
Practice	WB Exercises 23-24, pp. 61-62

Exercise 23

1. (a) 2.44   (b) 2.55

   (c) 0.07   (d) 8.78

   (e) 3.24   (f) 4.76

   (g) 6.15   (h) 5.43

61

Exercise 24

1. T: 2.35   E: 3.08
   H: 0.43   U: 4.65
   R: 4.67   P: 0.78
   C: 7.24   S: 1.37
   I: 7.38   G: 4.16
   O: 8.96   N: 6.78
   PENGUIN
   OSTRICH

62

2.1h Estimate Differences

Objectives

- Estimate differences for decimals.
- Use mental math strategies to add or subtract decimals close to a whole.

Note

The mental math strategies used here are similar to the ones that were used to add and subtract dollars and cents when the cents were close to a whole number of dollars in both *Primary Mathematics* 2B and 3B. If students have difficulty, use fraction squares to illustrate the process.

Estimation in subtracting decimals	
Write the expressions **0.83 − 0.29** and **8.3 − 0.29** on the board. Then write 0.54 on the board and ask students which of the two problems it is the correct answer to. Students should be able to make a quick estimate. Since 0.83 is less than 1 and we are only subtracting a number less than 1 from 8, obviously it is the answer to 0.83 − 0.29 and not 8.3 − 0.29. Remind students that estimation can help us quickly check whether an answer makes sense, particularly with regard to place values.	$0.83 − 0.29$ $8.3 − 0.29$ $0.54 \ ?$ $0.54 = 0.83 − 0.29$
Write the expression **62.1 − 4.83** on the board. Ask students to estimate the answer. Write down their estimates. Then have students find the exact answer. Tell them that we can round in whichever way allows us to make a quick estimate depending on how good we are at mental math and the situation. 57 is the estimate closest to the exact answer and we would use it in situations where we want to come as close as we can without doing the calculations. But to check if an answer is reasonable with respect to place value, even 60 is a good estimate, seeing that we are subtracting a number less than 10 from a number close to 62. That would tell us that 1.38 is not a likely answer, if we misplace the decimal, or 138 is not a reasonable answer, if we forget the decimal altogether.	$62.1 − 4.83$ Estimates: 57 55 $62.1 − 4.83 = 57.27$ 60 $62 − 5 = 57$ $60 − 5 = 55$ $\begin{array}{r}6.\ 2\ 1\\ -\ 4.\ 8\ 3\\ \hline 1.\ 3\ 8\end{array}$
Mental math strategies	
Write the expression **8.27 − 0.98** and ask students for an estimate. Tell them that since 0.98 is close to 1, 7.27 is a closer estimate. Ask them how we can use subtracting 1 to find the exact answer. If we do subtract 1, we have subtracted 2 hundredths too much, so we can just add that back in to get the exact answer of 7.29.	$8.27 − 0.98$ Estimates: 7 7.27 $8.27 − 0.98 = 8.27 − 1 + 0.02 = 7.29$
You can illustrate this with money. If you have $8.27 in your purse, and want to buy something that costs $0.98, what would you give the cashier? One of the dollars. You get 2 pennies change, and in all you have $7.29 left, one less dollar and two more cents.	

Write the expression **9.78 + 0.95** and ask students for an estimate. Include 10.78 and tell them this is a very close estimate. Ask them how much off from the exact answer it is: only 5 hundredths. Ask them if it is 5 hundredths too much or too little. It is 5 hundredths too much, since we added 1 to get the estimate. Ask them how we can use adding 1 to find the exact answer. Since we have added 5 hundredths too much, we can simply then subtract 5 hundredths.	$9.78 + 0.95$ Estimates: 11 10 10.78 $9.78 + 0.95 = 9.78 + 1 - 0.05 = 10.73$
Write the expression **40.63 − 3.97** and ask students to solve using mental math. 3.97 is close to 4, so they can subtract 4 and add back in the difference.	$40.63 - 3.97 = 40.63 - 4 + 0.03$ $= 36.63 + 0.03$ $= 36.66$
Provide other examples as needed, such as those at the right.	$6.29 + 4.99 = 11.29 - 0.01 = 11.28$ $9.97 + 34.2 = 44.2 - 0.03 = 44.17$ $9.45 - 4.95 = 4.45 + 0.05 = 4.5$
Assessment	**Text p. 40, Tasks 28-34**
	28. 20 29. (a) 15.87 (b) 50.01 (c) 39.57 30. (a) 2.63 (b) 21.89 (c) 27.54 31. 7.27 32. 9.98 33. 3.63 34. (a) 5.86 (b) 10.8 (c) 9.98 (d) 3.53 (e) 2.04 (f) 4.11
Mental Math Practice	Mental Math 16
Practice	WB Exercise 25, p. 63

Exercise 25

1. (a) 7.24; 7.23; 7.23
   (b) 11.63; 11.58; 11.58
   (c) 1.82; 1.83; 1.83
   (d) 4.05; 4.07; 4.07

2. (a) 9.79
   (b) 10.64

3. (a) 4.26
   (b) 4.58

63

Objectives

♦ Solve word problems involving addition or subtraction of decimals.

Note

The textbook shows the models already drawn. In order to give students practice in drawing models, you may want to write the problems on the board and guide students in drawing the models based on the information in the problem. Or, you can use the models in the textbook as examples of completed models, and have students practice drawing models using the Practice 2A problems instead. After students read the problem ask them what we need to find, whether we are given parts and asked to find a whole or a whole and asked to find a part, or whether we are comparing two numbers in some way.

Discussion	Text pp. 41-42, Tasks 35-37
Task 35: We are given three parts, the amount she spends on each item, and asked to find a whole, the total amount she spent, so the text shows a part-whole model.	35. $11.14 $11.14
Task 36: Since we are told how much longer one ribbon is than the other, the two lengths are being compared, so the text shows a comparison model and one solution where we find the length of the blue ribbon first and then add the two lengths. You can discuss an alternate solution. Since the blue ribbon is the same length as the white ribbon plus another part, we could also combine steps and simply write the expression: 1.85 m + 1.85 m + 1.4 m.	36. 5.1 5.1 m 37. $32.95 $32.95 $32.95 $32.95
Task 37: We are given the whole, how much she spent, and two parts, how much she spent on the fish and how much on the shrimp. We need to find a missing part.	
Assessment	
Select some problems from Practice 2A or 2B for students to work on. Have some present their solutions. See if there are any alternate methods or diagrams that other students might have, and compare the methods used. See answers on the next page of this guide.	

Enrichment

Write the word problem at the right on the board and discuss its solution.

Drawing a model for this problem can be tricky, because we do not know initially whether Abe or Ben should end up with a longer bar. If we at least get started on the model, though, we can see that an estimate will tell us whether Ben's bar should be shorter or longer than Abe's. The sum of what Abe saved and Ben spent is about $9 + $7, or $16. This is greater than the difference initially between the two bars. So that tells us that Ben's bar will be shorter than Abe's in the end.

Allow students to brainstorm on how to find the difference between the two final bars. We need to find the amount of overlap.

If students have difficulty, rephrase the problem using whole numbers.

Abe had $12 less than Ben. Abe saved another $9 and Ben spent $7. Who has less money now and how much less?

If students still have difficulty label the part of the bar that is the same for Abe and Ben by any amount, say 10, to verify the answer. Then, using the whole number version, Abe started with 10, saved 9 and ended up with 19. Ben started with 22, spent 7, and ended up with 15. Ben has 4 less than Abe. 12 is 4 less than 9 and 7.

Point out that simplifying a problem, such as solving a similar problem using whole numbers, can help lead to a solution.

Abe had $12.07 less than Ben. Abe saved another $8.95 and Ben spent $7.29. Who has less money now and how much less?

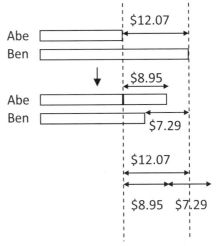

$8.95 + $7.29 = $16.24
$16.24 − $12.07 = $4.17

Ben has $4.17 less than Abe.

Practice

WB Exercises 26-27, pp. 64-66

Exercise 26

1. 5 yd − 2.35 yd = **2.65 yd**
   He used 2.65 yd of wire.

2. 5 kg − 3.6 kg = **1.4 kg**
   He gained 1.4 kg.

3. $36.45 − $2.54 = **$33.91**
   He spent $33.91.

64

Exercise 27

1. Amount spent:
   $1.40 + $2.50 = $3.90
   Amount left:
   $13.50 − $3.90 = **$9.60**
   She had $9.60 left.

2. Cost of items:
   $12 + $4.50 = $16.50
   Change:
   $20 − $16.50 = **$3.50**
   She received $3.50 in change.

65

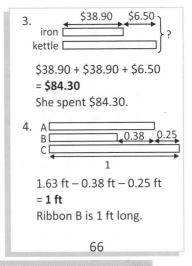

3.
$38.90 + $38.90 + $6.50
= **$84.30**
She spent $84.30.

4.
1.63 ft − 0.38 ft − 0.25 ft
= **1 ft**
Ribbon B is 1 ft long.

66

Objectives

♦ Practice.

Note

Do not require students to draw models if they don't need to. Rather than forcing a student to draw a model when he or she can solve the problem easily without one, provide the student with some more challenging problems from one of the supplementary books where a model is needed. If a student struggles, though, encourage him or her to draw a model for the problem.

Practice	Text p. 43, Practice 2A
	1. (a) 0.9　　(b) 1.7　　(c) 4.1 2. (a) 0.1　　(b) 0.11　　(c) 1.26 3. (a) 0.1　　(b) 1.6　　(c) 2.6 4. (a) 0.03　　(b) 0.93　　(c) 3.35 5. (a) 8.3　　(b) 0.82　　(c) 2.02 6. (a) 2.5　　(b) 0.84　　(c) 0.87 7. (a) 8.85　　(b) 7.58　　(c) 20.04 　　(d) 2.8　　(e) 4.51　　(f) 3.64 8. 1.32 m − 0.07 m = **1.25 m** Brianne is 1.25 m tall. 9. $5.75 + $7.50 = **$13.25** She spent $13.25 on meat. 10. $16.80 + $3.60 = **$20.40** He had $20.40 at first. 11. 42.5 kg − 38.6 kg = **3.9 kg** He lost 3.9 kg. 12. 15.3 s − 14.5 s = **0.8 s** **Fred** ran 0.8 s faster.

Practice	Text p. 44, Practice 2B
	1. (a) 48.68 (b) 19.43 (c) 40.02

Practice | **Text p. 44, Practice 2B**

1. (a) 48.68   (b) 19.43   (c) 40.02
2. (a) 28.6   (b) 17.31   (c) 19.98
3. (a) 13.33   (b) 22.23   (c) 4.89
4. (a) 36.65   (b) 11.05   (c) 10.61

5. 1.69 lb + 1.69 lb + 2.51 lb = **5.89 lb**
   The total weight is 5.89 lb.

6. 3 ℓ − 0.5 ℓ − 0.25 ℓ = **2.25 ℓ**
   She had 2.25 ℓ of milk left.

7.
   5.85 km − 1.7 km = 4.15 km
   5.85 km + 4.15 km = **10 km**
   He jogged a total of 10 km.

8. 24.8 cm + 12.6 cm + 18.4 cm = **55.8 cm**
   She had 55.8 cm of ribbon at first.

9. $4.15 + $6.80 = $10.95
   $15 − $10.95 = **$4.05**
   Lucy spent $4.05.

10. $4.90 + $7.50 = $12.40
    $15 − $12.40 = **$2.60**
    She received $2.60 change.

2.2 Multiplication

Objectives

♦ Multiply 1-place and 2-place decimals by a whole number.
♦ Estimate the product.
♦ Solve word problems involving multiplication of decimals.

Material

♦ Place-value discs

Prerequisites

Students should know multiplication facts through 10 x 10 thoroughly. They should be able to multiply whole numbers of at least 4 digits by 1 digit using the standard algorithm, and be comfortable with place-value concepts.

Notes

In *Primary Mathematics* 3A students learned the standard algorithm for multiplying a whole number by a 1-digit whole number. In this part the standard algorithm will be extended to 1-place and 2-place decimals.

When writing addition or subtraction problems in vertical format, it is important to always align the digits according to their place value. In multiplication, however, we do not align digits for the two factors, since we need to multiply the single digit with each of the digits in the decimal. While it is possible to represent 6.14 x 3 in either of the two ways shown on the right, only the second way will be used. In *Primary Mathematics* 5A students will learn how to multiply a decimal by a whole number greater than 10 or another decimal, without regard to the decimal point when performing the actual multiplication, and then place the decimal point correctly in the final answer.

```
    6 . 1 4
  x     3
  1 8 . 4 2


    6 . 1 4
  x         3
  1 8 . 4 2
```

As with addition and subtraction, always use the place values when discussing the process. 0.6 x 5 is "6 tenths times 5 equals 30 tenths or 3 ones," not "six times five is three."

Students will be asked to estimate their answers in order to check if their answers are reasonable. For multiplication, they can round the decimal to one non-zero digit, e.g., 27.9 x 3 can be estimated by using 30 x 3 = 90 and 0.279 x 3 can be estimated using 0.3 x 3 = 0.9.

In *Primary Mathematics* 3, students learned to apply the part-whole and comparison models for problem-solving situations involving multiplication or division with whole numbers. Here they will extend their use of these models to involve decimal numbers.

In a part-whole model equal parts are drawn as equal units.

For example, we are told that a shirt costs $10.49 and asked for the cost of 5 shirts. Given the value of a unit ($10.49) and the number of units (5 shirts), we can see from the model that to find the total we multiply.

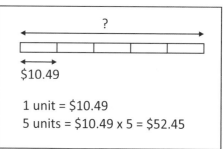

1 unit = $10.49
5 units = $10.49 x 5 = $52.45

In a comparison model one value is some multiple of another.

For example, we could be told that a jacket costs 3 times *more than* a shirt and asked to find the total cost of the jacket and shirt, given the cost of the shirt. We can draw one unit for the cost of the shirt, 4 units for the jacket, and the total cost is 5 units. From the model, we can see that we multiply to find the total cost.

These two basic types of models can be combined to illustrate more complicated problems.

For example, if we are told that a shirt costs $10.49, and someone bought 4 shirts and a pair of shoes and spent a total of $72.21, we could model one part with 4 units for the shirts, and another part for the shoes, and show the total. We can see from the model that we find the cost of the shoes by first multiplying to get the cost of all the shirts, and then subtracting that part from the total to find the cost of the shoes.

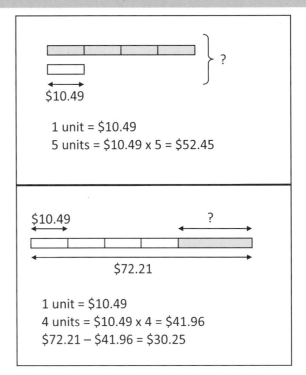

1 unit = $10.49
5 units = $10.49 x 5 = $52.45

1 unit = $10.49
4 units = $10.49 x 4 = $41.96
$72.21 − $41.96 = $30.25

Students should be proficient at drawing models when needed but need not be required to draw a model for every exercise problem if they can solve the problem correctly without a drawing.

Do not insist that students follow a prescribed set of steps in drawing the models. Unless they learn to solve these problems logically and develop problem-solving skills that don't rely on following a set of steps, they will have difficulty with more complex problems that don't fit a specific set of steps.

2.2a Multiply Tenths and Hundredths

Objectives

♦ Multiply tenths.
♦ Multiply hundredths.

Note

Multiplication of a decimal by a whole number is somewhat more challenging than adding or subtracting decimal numbers. Illustrate the process with place-value discs as much as needed.

Discussion	Text p. 45
The top half of p. 45 in the text illustrates multiplying tenths using measurement as the concrete introduction. If we have 0.4 liters in each of 3 separate beakers, then the total amount is 1.2 liters. We have multiplied 4 tenths by 3 to get 12 tenths, which is the same as 1.2. If you have beakers or jars marked in tenths and can use colored water in your classroom, you can demonstrate this concept more concretely.	0.4 x 3 = **1.2** She drinks **1.2** liters of milk in 3 days. 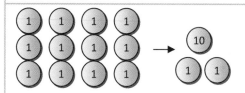 4 ones x 3 = 12 ones 4 x 3 = 12
The second half of the page shows the same problem with place-value discs, which is a more abstract illustration, and then pictorially with a bar model. Display or draw 3 rows of 4 1-discs. Write the equations as shown at the right.	
Repeat with 0.1-discs. Point out that each disc now represents a tenth. Write the equations as shown, and then vertically. Remind students that 12 tenths is the same as 1 whole and 2 tenths. The multiplication fact we use to find the answer is the same as with ones, but we are multiplying tenths rather than ones and the answer is in tenths.	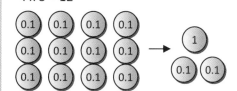 4 tenths x 3 = 12 tenths 0 . 4 0.4 x 3 = 1.2 x 3 1 . 2

Tell students that when we rewrite the problem vertically to multiply decimals, rather than aligning the digits by place-value, we align them on the right. "x 3" is the same as adding three 0.4's together. When we are multiplying a decimal by a whole number, the place value of the answer and the position of the decimal depends only on the decimal number.	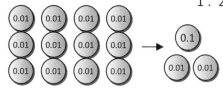
	4 hundredths x 3 = 12 hundredths
	0.04 x 3 = 0.12
	       0 . 0 4
	    x 3
	     0 . 1 2

Repeat with 0.01 discs. Point out that the number of decimal places in the answer is the same as in the decimal number we are multiplying.	
Repeat the process again for 5 x 4, 0.5 x 4, and 0.05 x 4, using discs. This time, the answer may not have the same number of decimal places as 0.5 or 0.05. Point out that extra 0's after the last non-zero decimal place can be dropped without changing the value of the answer. We still use the math fact for 4 x 5 = 20.	5 ones x 4 = 20 ones 5 x 4 = 20 5 tenths x 4 = 20 tenths 0.5 x 4 = 2.0 = 2 5 hundredths x 4 = 20 hundredths 0.05 x 4 = 0.20 = 0.2

Discussion	Text pp. 46-47, Tasks 1-3
Task 1: Just as 2 ones x 4 is 8 ones, so 2 tenths x 4 is 8 tenths and 2 hundredths x 4 is 8 hundredths.	1. (a) 0.8 (b) 0.08
Tasks 2-3: After we multiply tenths or hundredths, using the same math facts for multiplying ones, we rename 10 tenths as 1 one and 10 hundredths as 1 tenth. You can show the process with place-value discs if needed. For 2(b) and 3(b) renaming results in the 0 of 30 no longer being needed in the answer, since it is after the last non-zero digit after the decimal.	2. (a) 2.1 (b) 3 3. (a) 0.21 (b) 0.3
Write the problems on the right and have students supply the answers. The same math fact is used regardless of the place value of the digit. Be sure to use the place-value names when discussing these problems.	$\underline{7}00$ x $\underline{3}$ = $\underline{21}00$ $\underline{7}0$ x $\underline{3}$ = $\underline{21}0$ $\underline{7}$ x $\underline{3}$ = $\underline{21}$ 0.$\underline{7}$ x $\underline{3}$ = $\underline{2}$.$\underline{1}$ 0.0$\underline{7}$ x $\underline{3}$ = 0.$\underline{21}$ $\underline{6}00$ x $\underline{5}$ = $\underline{3000}$ $\underline{6}0$ x $\underline{5}$ = $\underline{300}$ $\underline{6}$ x $\underline{5}$ = $\underline{30}$ 0.$\underline{6}$ x $\underline{5}$ = $\underline{3.0}$ = 3 0.0$\underline{6}$ x $\underline{5}$ = 0.$\underline{30}$ = 0.3
Assessment	**Text p. 47, Tasks 4-6**
Students should be able to do these mentally; it is not necessary to ask them to rewrite the problems vertically here.	4. (a) 6 (b) 0.6 (c) 0.06 (d) 28 (e) 2.8 (f) 0.28 (g) 40 (h) 4 (i) 0.4 5. $3.20 6. (a) $0.80 (b) $4.20 (c) $7.20
Mental math practice	Mental Math 17
Practice	WB Exercise 28, pp. 67-68

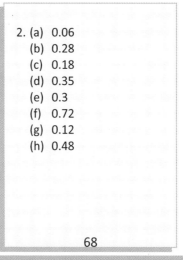

Exercise 28

1. (a) 0.8
   (b) 1.8
   (c) 1.4
   (d) 3.6
   (e) 3
   (f) 5.6
   (g) 2.7
   (h) 4

2. (a) 0.06
   (b) 0.28
   (c) 0.18
   (d) 0.35
   (e) 0.3
   (f) 0.72
   (g) 0.12
   (h) 0.48

67

68

2.2b Multiply Decimals

Objectives

♦ Multiply 1-place decimals by a whole number.

Note

Take what time is needed to first make sure students understand the process with 1-place decimals before going immediately on to 2-place decimals. For slower students use actual place-value discs as needed, rather than just having students look at the static pictures in the textbook. For more capable students the pictures may be sufficient, and you may be able to combine this lesson with the next one.

Multiply 1-place decimals	
Write the expression **4.6 x 3** on the board. Remind students that 4.6 x 3 is the same as 4.6 + 4.6 + 4.6. So we can simply multiply 0.6 by 3, rename tenths if we have too many, and then multiply 4 by 3. Illustrate the steps with place-value discs if needed, setting out 4 ones and 6 tenths, tripling the tenths, replacing 10 tenths with a one, tripling the original 4 ones, and adding in the renamed one. Point out that we worked the problem in exactly the same way we would work the problem 46 x 3, but instead of multiplying 46 ones, we are multiplying 46 tenths. The answer is therefore in tenths. 138 tenths is 13.8.	4.6×3

4.6×3

$$\begin{array}{r} 1 \\ 4\;.\;6 \\ 4\;.\;6 \\ +\;\;4\;.\;6 \\ \hline 1\;3\;.\;8 \end{array}$$

$$\begin{array}{l} 4.6 \times 3 \\ \wedge \\ 4\;\;\;0.3 \\ 0.6 \times 3 = \quad 1\;.\;8 \\ 4 \times 3 = \quad \underline{1\;2} \\ \qquad\qquad\;\; 1\;3\;.\;8 \end{array}$$

$$\begin{array}{r} 4\;.\;6 \\ \times\qquad 3 \\ \hline \end{array} \rightarrow \begin{array}{r} 1 \\ 4\;.\;6 \\ \times\qquad 3 \\ \hline .\;8 \end{array} \rightarrow \begin{array}{r} 1 \\ 4\;.\;6 \\ \times\qquad 3 \\ \hline 1\;3\;.\;8 \end{array}$$

$0.6 \times 3 = 1.8 \qquad 4 \times 3 = 12$

$1.8 + 12 = 13.8$

Repeat with 20.7 x 6, discussing each step and illustrating with place-value discs if needed.

$$\begin{array}{r} 4 \\ 2\;0\;.\;7 \\ \times\qquad 6 \\ \hline .\;2 \end{array} \quad (0.7 \times 6 = 4.2)$$

$$\downarrow$$

$$\begin{array}{r} 4 \\ 2\;0\;.\;7 \\ \times\qquad 6 \\ \hline 4\;.\;2 \end{array} \quad (0 \times 6 + 4 = 4)$$

$$\downarrow$$

$$\begin{array}{r} 4 \\ 2\;0\;.\;7 \\ \times\qquad 6 \\ \hline 1\,2\,4\;.\;2 \end{array} \quad (20 \times 6 = 120)$$

Discuss the steps for 82.5 x 8. Point out that in the final answer we can drop the 0 after the decimal point, but not the 0 before the decimal point. The 0 before the decimal point is needed to show that the 6 is in the tens place, not the ones place, but the 0 after the decimal point is not needed to keep any digit in place.	$$\begin{array}{r}\overset{4}{}8\,2\,.\,5\\ x\qquad 8\\ \hline .\,0\end{array}\quad(0.5 \times 8 = 4.0)$$ ↓ $$\begin{array}{r}\overset{2\ 4}{8\,2\,.\,5}\\ x\qquad 8\\ \hline 0\,.\,0\end{array}\quad(2 \times 8 + 4 = 20)$$ ↓ $$\begin{array}{r}\overset{2\ 4}{8\,2\,.\,5}\\ x\qquad 8\\ \hline 6\,6\,0\,.\,0\end{array}\quad(80 \times 8 + 2 = 660)$$ 82.5 x 8 = 660

Assessment	
Write some problems horizontally on the board and have students rewrite them vertically and find the answers.	8.9 x 2 17.8 5.5 x 7 38.5 2.5 x 8 20 58.3 x 6 349.8 9 x 13.6 122.4 88.8 x 8 710.4 4 x 890.5 3562

Practice	
	WB Exercise 29, p. 69

Exercise 29

1. (a) 8.6   (b) 19.2

   (c) 16.8   (d) 42.3

   (e) 27.6   (f) 38.5

   (g) 132.5   (h) 244.8

69

Objectives

♦ Multiply 2-place decimals by a whole number.
♦ Estimate the product.

Note

Once students are comfortable with multiplying by a 1-place decimal they should be able to easily extend the process to additional decimal places. Show the steps with place-value discs as needed.

Multiply 2-place decimals, estimate	Text p. 48, Tasks 7-8
Task 7: Rewrite the problem on the board and discuss the steps in solving it, using place-value discs to illustrate if needed. Be sure to use place-value language. We are multiplying 5 hundredths by 3, not 5 by 3, in the first step. Note that we can wait until after we multiply tenths before writing the decimal point, but we need to be sure that 5 is in the hundredths place with respect to the decimal number.	7. 0.75 8. 9.06

Task 7 worked example:

$$\begin{array}{r} \overset{1}{} \\ 0\,.\,2\,5 \\ \times\qquad 3 \\ \hline 5 \end{array}$$ (0.05 x 3 = 0.15)

↓

$$\begin{array}{r} \overset{1}{} \\ 0\,.\,2\,5 \\ \times\qquad 3 \\ \hline 0\,.\,7\,5 \end{array}$$ (0.2 x 3 + 0.1 = 0.7)

Ask students if they could do this problem mentally. They can think of 2-place decimals in terms of money. $0.25 is the same as a quarter. Ask them what the value of 3 quarters is.

Task 8: Again, write the problem on the board and discuss the steps in solving it, or have a student come to the board and write the steps.

Remind students that we can estimate the answer to multiplication problems as well as addition and subtraction problems. For multiplication problems, the easiest estimate to make is to round both numbers to a single non-zero digit. For these problems, where one of the factors is already a single-digit whole number, we only have to round the decimal number. Ask students what they would round 4.53 to. Since it is slightly closer to 5, we would round it to 5 and the estimated answer would be 10. The actual answer of 9.06 is close to 10. Estimation will help prevent errors in place value. We know from the estimate that an answer of 906, which we would get by forgetting to put the decimal point in, does not make sense. Neither would 90.6.

Task 8 worked example:

$$\begin{array}{r} 4\,.\,5\,3 \\ \times\qquad 2 \\ \hline 6 \end{array}$$ (0.03 x 2 = 0.06)

↓

$$\begin{array}{r} \overset{1}{} \\ 4\,.\,5\,3 \\ \times\qquad 2 \\ \hline 0\,6 \end{array}$$ (0.5 x 2 = 1.0)

↓

$$\begin{array}{r} \overset{1}{} \\ 4\,.\,5\,3 \\ \times\qquad 2 \\ \hline 9\,.\,0\,6 \end{array}$$ (4 x 2 + 1 = 9)

4.53 x 2
5 x 2 = 10
The answer will be close to 10.

Refer back to Task 7. Ask students how they would estimate the answer to this problem. We could use 0.2 x 3 = 0.6, but then we have to be sure to remember that the answer is in tenths. So in order not to make place-value errors in estimates, it helps if we can avoid decimals altogether if possible. For decimals less than 1, the answer will always be less than the number we are multiplying by. So only an answer less than 3 will make sense.

0.25 x 3
1 x 3 = 3
The answer will be less than 3.

Ask students to estimate the answer to 0.48 x 8. One possible estimate is 0.5 x 8 = 4. Another valid estimate is that 0.48 is about half, so the answer will be about half of 8. Without any multiplication, we also know the answer will be less than 8.	0.48 x 8 = ? 0.5 x 8 = 4: The answer will be about 4. The answer will be half of 8. The answer will be less than 8. 0.48 x 8 = 3.84
Discussion	**Text p. 49, Task 10**

You can have students do these problems independently first, and then discuss the steps, or you can discuss the steps as a class, perhaps asking a student to explain the steps at the board. Ask students to first estimate the answer.	Estimate:
	10. (a) 124.2 20 x 6 = 120
	    (b) 260.8 30 x 8 = 240
	    (c) 414.09 50 x 9 = 450
	    (d) 180.81 30 x 7 = 210

Assessment	**Text p. 49, Tasks 9, 11-14**

For Task 14 make sure that students realize that when writing the answer to a problem involving money in decimal notation, the 0 needs to be kept in the hundredths place, since the cents after the decimal point are always written to 2 places. If there are no cents at all, then the two 0's for cents can be dropped if desired.	9. (a) 12.9 (b) 1.04 (c) 12.48
	   (d) 11.8 (e) 2.25 (f) 36.16
	
	11. (a) 37 (b) 102.06 (c) 289.56
	    (d) 80.4 (e) 180.75 (f) 442
	
	12. 63
	
	13. (a) 19.5 (b) 5.76 (c) 178.38
	
	14. (a) $8.20 (b) $117 (c) $292.05

Enrichment - Mental Math	Mental Math 18
Practice	WB Exercises 30-31, pp. 70-71

Exercise 30

1. (a) 1.66   (b) 0.72

   (c) 15.78   (d) 27

   (e) 42.18   (f) 45.12

   (g) 579.46 (h) 582.48

70

Exercise 31

1. L: 0.96   H: 81.2
      E: 0.21   Y: 14.73

   T: 32.25   E: 561
      P: 726.3   E: 64.44

   N: 36.45   D: 3265.6
      H: 28.94   E: 78.48

   HELP THE NEEDY

71

2.2d   Word Problems

Objectives

♦ Solve word problems involving multiplication of decimals.

Note

In *Primary Mathematics* 3B the bar models for multiplication and division consisted of equal *units*. Use the idea of units here. Write the equations to include the number of units. This provides a good beginning to the idea of balancing equations. For example, if the number of units goes from 1 unit to 4 units on one side, then the value of one unit is multiplied by 4 on the other side.

Part-whole and comparison models for multiplication	Text p. 50, Tasks 15-16
Write the problems on the board and relate the bar models to the information in the problems.	15. $23.70 $23.70
Task 15: Ask students if we are comparing two numbers. We are not. Then ask them whether the problem asks for a part or a whole. It asks for a whole, so we will likely either use addition or multiplication. Ask them if there is more than one part and if so, whether they are equal. There are 6 equal parts.	16. $80.20 $80.20
Each part in the model is called a unit. Point out that each unit in the model is the same. Even if our drawing is not as nice as the ones in the book, when we draw it we know that each little rectangle of about the same length is meant to be the same as we draw it. When we are given equal units and the value of 1 unit, we can find the value of multiple units using multiplication.	
Write the equations at the right for Task 15, having students supply the answers.	15. 1 unit = $3.95 6 units = $3.95 x 6 = $23.70
Task 16: Ask students whether we are comparing two numbers. We are, the amounts Rachel and Susan saved. The problem tells us that Susan saved 4 times as much as Rachel, so the amount Rachel saved is 1 unit, and the amount Susan saved is 4 units. Write the equations at the right for Task 16, getting students to supply the answers.	16. 1 unit = $20.05 4 units = $20.05 x 4 = $80.20
Ask students to also find out how much Susan and Rachel saved altogether. The total number of units is 5. If we had not first been asked to find out how much Susan saved, we could find the amount they both saved by multiplying the value of 1 unit by 5.	5 units = $20.05 x 5 = $100.25 They saved $100.25 altogether.
Then ask students to find out how much more Rachel saved than Susan using multiplication only. We can multiply the value of 1 unit by 3.	3 units = $20.05 x 3 = $60.15 Susan saved $60.15 more than Rachel.

Discussion	Text p. 51, Tasks 17-18

You can ask students to draw models for these two tasks. Some suggested models are shown here.
 	
Task 17: In this task we are given a whole, the amount Sam had. He spent some money and had some left, and we are asked to find the part he had left. So we can draw a part-whole model. One part is the 4 sets of stamps, so we divide it into 4 equal units. The first step for this problem is to find the value of one part using multiplication. Then we can find the missing part using subtraction.
 	
Task 18: In this task we need to find a whole, the total amount of material she bought. One part is the amount she has left after making the curtains and the other part is the amount used to make the curtains. Since we are told there were 6 curtains and 3.15 m was used for each, this part consists of 6 equal units. The first step for this problem is again to find the value of one part using multiplication. Then we can find the whole using addition.
	17. $18.20
	    $18.20
	
	18. 21.4
	    21.4 m

	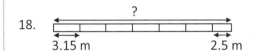

Assessment	
Write the problem at the right on the board and have students draw a model and solve it. You can also select some problems from Practice 2C on p. 52 of the textbook.	Rope A is three times as long as Rope B. Rope C is 5.7 meters longer than Rope A. Rope B is 4.45 meters long. How long are all three ropes together? 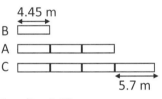 1 unit = 4.45 m 7 units = 4.45 m x 7 = 31.15 m 31.15 m + 5.7 m = 36.85 m Altogether, the ropes are 36.85 m long.

Practice	WB Exercises 32-33, pp. 72-74

Exercise 32	Exercise 33	

1. 1 unit = 1.25 yd
   3 units = 1.25 yd x 3 = **3.75 yd**
   The total length is 3.75 yd.
 	
2. 1 unit = 5.7 ℓ
   5 units = 5.7 ℓ x 5 = **28.5 ℓ**
   The capacity of the fish tank is 28.5 ℓ.
 	
3. 1 unit = $2.50
   6 units = $2.50 x 6 = **$15**
   He saved $15 altogether.
	1. $6.90
	$1.45 x 2 = $2.90
	Total: $9.80
	
	$3.75 x 2 = $7.50
	$0.95 x 2 = $1.90
	Total: $9.40
	
	          $9.95
	$1.20 x 4 = $4.80
	Total: $14.75
	
	$8 x 3 = $24.00
	          $16.50
	Total: $40.50
		2.
		5 m / 0.85 / ?
		0.85 m x 2 = 1.7 m
		5 m – 1.7 m = **3.3 m**
		She had 3.3 m of material left.
		
		3.
		? / $1.35 / $2.50
		$1.35 x 6 = $8.10
		$8.10 + $2.50 = **$10.60**
		She had $10.60 at first.

72	73	74

Objectives

♦ Practice.

Note

Have students draw models if needed. The need will vary with students. Encourage them to include the term **unit** in their equations.

Practice	Text p. 52, Practice 2A
	1. (a) 8.6 (b) 11 (c) 2.28

Text p. 52, Practice 2A

1. (a) 8.6   (b) 11   (c) 2.28
2. (a) 6.23   (b) 4.52   (c) 4.14
3. (a) 5.3   (b) 3.5   (c) 1.85
4. (a) 1.52   (b) 3.17   (c) 2.85
5. (a) 3.6   (b) 5.6   (c) 1.86
6. (a) 1.35   (b) 6   (c) 19.3

7. (a) 18; 19.2   (b) 6; 7.44   (c) 20; 20.45

8. 1.5 m − 1.39 m = **0.11 m**
   The difference between the two results is 0.11 m.

9.

   1 unit = 13.45 lb
   3 units = 13.45 lb x 3 = **40.35 lb**
   He used 40.35 lb of sand.

10. 1 packet spice: $0.85
    4 packets spice: $0.85 x 4 = $3.40
    $3.40 + $3.75 = **$7.15**
    She spent $7.15.

11. (1.46 ℓ + 0.8 ℓ) − 0.96 ℓ = 2.26 ℓ − 0.96 ℓ = **1.3 ℓ**
    He had 1.3 ℓ of gray paint left.

12.

    1 unit = $2.35
    5 units = $2.35 x 5 = $11.75
    $20 − $11.75 = **$8.25**
    She received $8.25 change.

Enrichment

Write the problem at the right on the board and give students a chance to work on it on their own. There is more than one way to solve the problem. One method is suggested below. You can allow students to discuss their methods.	If Amy gives $3.65 to Zoe, she will have three times as much money as Zoe. If Amy gives $7.20 to Zoe, she will have twice as much money as Zoe. How much money does Amy have?

In both cases the total amount stays the same. So draw two bars that are the same for each case.

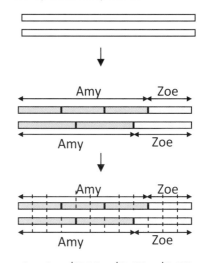

In the first case Amy ends up with 3 times as much as Zoe, so there are 4 total units (3 for Amy, one for Zoe). In the second case there are 3 equal units.

In order to compare the two bars subdivide them so that there are equal units. Now we can see that the difference in what she gave Zoe in the two cases is 1 of these smaller units

Using the first case, Amy ends up with 9 units.

1 unit = $7.20 − $3.65 = $3.55
9 units = $3.55 x 9 = $31.95
She started with $3.65 more.
$31.95 + $3.65 = $35.60
Amy had $35.60.

2.3 Division

Objectives

♦ Divide decimal numbers of up to two places by a 1-digit whole number.
♦ Estimate the quotient for division of decimals.
♦ Round the quotient to one decimal place.
♦ Solve word problems involving division of decimals.

Material

♦ Place-value discs

Prerequisites

Students should know the division facts that correspond to multiplication facts through 10 x 10 thoroughly. They should be able to divide whole numbers of at least 4 digits by 1 digit using the standard algorithm, and be comfortable with place-value concepts.

Notes

In *Primary Mathematics* 3A students learned the standard algorithm for dividing a whole number by a 1-digit whole number. In this part the formal algorithm for division is extended to dividing a decimal number by a whole number. In *Primary Mathematics* 5A students will learn how to divide by a decimal.

The dividend is the quantity to be divided and the divisor is the quantity by which another quantity is to be divided. Because it is far more important for students to learn division than to learn the formal names of the process, these terms are not used in the course material itself and will be used only in the guide to avoid having to use terms such as "number being divided."

$$\text{dividend} \div \text{divisor} = \text{quotient}$$
$$\text{divisor} \overline{)\text{dividend}}^{\text{quotient}}$$

In previous units when a whole number was divided by another whole number with a remainder of ones, students expressed the answer in the form of a whole number quotient with a remainder ($29 \div 4 = 7$ R 1). In this part students will learn to find a decimal quotient and to round decimal quotients to 1 decimal place. ($29 \div 4 = 7.3$, rounded to 1 decimal place.)

Students will be finding an estimate to check the reasonableness of the actual answer. In division rather than rounding the dividend to the nearest whole number or ten or other multiple of ten, we round to the closest multiple of the divisor. For example, to estimate the answer for the division problem $31.2 \div 8$, it would not be any easier to divide 30 by 8 than it is to divide 31.2 by 8, so it is not helpful to round the dividend to the nearest ten in this case. 8 x 3 is 24 and 8 x 4 is 32; 31.2 is closer to 32 than 24, so $31.2 \div 8$ is about 4.

In *Primary Mathematics* 3 students learned to apply the part-whole and comparison models for problem-solving situations involving multiplication or division with whole numbers. Here they will extend their use of these models to involve decimal numbers.

The models used for division are the same as used for multiplication; a part-whole model, a comparison model, or some combination of the two. Instead of being given the value of a part, though, the value of a whole or of multiple parts is given, and the student needs to find the value of one part or a different number of parts, after first finding the value of one part.

For example, if 5 shirts cost $52.45, and we are asked for the cost of 2 shirts, we first find the cost of 1 shirt.

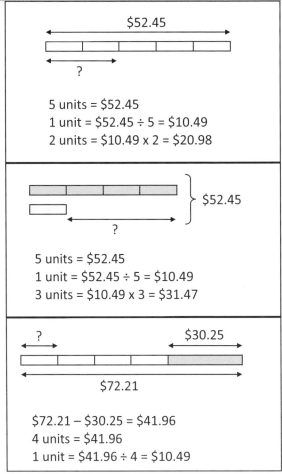

$52.45

5 units = $52.45
1 unit = $52.45 ÷ 5 = $10.49
2 units = $10.49 x 2 = $20.98

Or, if we are told that a pair of pants cost 4 times as much as a shirt and given the total cost of both pants and shirt, we can find the difference in cost by first finding the value of 1 unit.

$52.45

?

5 units = $52.45
1 unit = $52.45 ÷ 5 = $10.49
3 units = $10.49 x 3 = $31.47

Problems can involve 2 parts, where one part is divided into equal units. For example: 4 shirts and a pair of shoes cost $72.21. If the shoes cost $30.25, find the cost of one shirt.

? $30.25

$72.21

$72.21 – $30.25 = $41.96
4 units = $41.96
1 unit = $41.96 ÷ 4 = $10.49

In this guide there are a series of similar enrichment problems in the lessons that go along with the practices and reviews. An approach based on the information in the problem can give an easier and quicker solution than an approach based on steps that work with different but similar problems. The more practice students get in solving different types of problems without being given a set of specific steps, or looking at a solution someone else came up with, the more they will develop the problem-solving skills needed to solve any kind of math reasoning problem. Imposing a set of steps, while possibly making model drawing easier initially for struggling students, inhibits the development of these skills.

Model drawing is a powerful tool for solving word problems. It should be kept in proper perspective, though. The purpose for solving a problem is to find the answer, not to draw a model. Students at this level should be capable of using model drawing effectively and efficiently without being required to follow strict rules regarding type of model or placement of labels or question marks. The more practice they have with a variety of problems, the better they will become both at drawing effective models and at solving the problems.

2.3a Divide Decimals

Objectives

♦ Divide decimals using division facts.

Note

In this lesson all quotients will have a single non-zero digit. Thus the quotient will be determined using the division facts that correspond with multiplication facts for 10 x 10. The emphasis is on placing the decimal correctly in the quotient.

Discussion	Text p. 54
The top half of page 53 illustrates dividing tenths using measurement as the concrete introduction. If we divide 0.9 liters into 3 equal parts, each part has 0.3 liters. If you have beakers or jars marked in tenths and can use colored water you can demonstrate this. The second half of the page shows the same problem with place-value discs, and then pictorially with a bar model. Show or draw nine 1-discs, divide them into 3 groups, and write the division equation both horizontally and vertically. For the vertical representation, remind students that we write the amount in each group above the line, the amount put into groups below the number we are dividing, and the difference is the remainder. Repeat with nine 0.1-discs. Point out that we use the same division fact, 9 ÷ 3 = 3, but now we are dividing tenths. Point out that in the vertical working, we line up the quotient, the amount in each group, and the remainder with the number in the "box" that we are dividing (the divisor). Since the amount that goes in each group and the remainder are in the correct position, we don't need to include the decimal point. Repeat with nine 0.01-discs.	$0.9 \div 3 = \mathbf{0.3}$ There was **0.3** liter of water in each beaker. 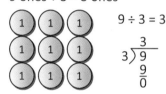
Set out a 1-disc and five 0.1-discs and ask what number this represents. Then ask how we could use the discs to divide by 3. We need to rename the one as tenths. 1.5 = 15 tenths. Then we can divide. Point out that we can simply recognize that 15 can be divided by 3, remember that 1.5 is the same as 15 tenths, and use the knowledge that 15 ÷ 3 = 5.	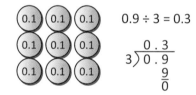
Now set out two 1-discs and ask students how we can divide this by 5. We have to rename both 1-discs as tenths to divide. When we see a written problem such as 2 ÷ 5, we can add a decimal point and 0's.	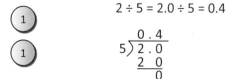

Discussion	Text pp. 54-55, Tasks 1-3
Write the problems on the board and illustrate the solutions with place-value discs, or have students use place-value discs and explain the process. Task 1: We can divide 6 tenths or 6 hundredths by 2 in the same way as we divide 6 ones by 2, but the answer is in tenths or hundredths. Task 2: We divide decimals the same way as we divide whole numbers when using the division algorithm. We start by trying to divide the number in the highest place, which in this case is 1. Since 1 whole cannot be divided by 3 we have to rename it as 10 tenths and add that to the 8 tenths we already have. We can then divide 18 tenths by 3. We use the division fact $18 \div 3 = 6$ to find the quotient. The quotient is 6 tenths. Similarly, to divide 2 by 4, we need to rename 2 as 20 tenths and divide that by 4. Show how you would do the problem vertically by adding a 0 after the decimal point and writing the answer above the line in the correct place, which is above the 0. Task 3: We can follow the same process with hundredths, renaming tenths as hundredths as necessary in order to divide.	1. (a) 0.3 (b) 0.03 2. (a) 0.6 (b) 0.5 3. (a) 0.06 (b) 0.05
Assessment	**Text p. 55, Tasks 4-6**

	4. (a) 2 (b) 0.2 (c) 0.02
	    (d) 5 (e) 0.5 (f) 0.05
	    (g) 6 (h) 0.6 (i) 0.06
	
	5. $0.70
	
	6. (a) $0.30 (b) $0.30 (c) $0.60

Mental Math Practice	Mental Math 19
Practice	WB Exercise 34, pp. 75-76

Exercise 34

1. (a) 0.4
   (b) 0.3
   (c) 0.3
   (d) 0.4
   (e) 0.4
   (f) 0.6
   (g) $0.70
   (h) $0.60

75

2. (a) 0.06
   (b) 0.05
   (c) 0.04
   (d) 0.06
   (e) 0.06
   (f) 0.06
   (g) $0.09
   (h) $0.05

76

2.3b Divide Hundredths

Objectives

♦ Divide 2-place decimals by a 1-digit whole number where the quotient is less than 1.

Note

In this lesson all quotients will be less than 1. Go through the division process using place-value discs, and allow students to use the discs as needed. More capable students will not need to use physical discs and you can just talk through the process.

Illustrate the division algorithm	
Provide students with place-value discs and get them to divide them up and rename as needed. On the board relate what they are doing step-by-step with the division algorithm. Write **1.11 ÷ 3** on the board and have students pick out 1 one, 1 tenth, and 1 hundredth. Ask students to first divide the ones by 3. 1 one cannot be divided into 3 groups. Show students that we write a 0 above the line in the ones place. Ask them how we can divide the one. They need to replace the one with 10 tenths. For the written algorithm tell them that we simply know that we now will divide 11 tenths. Point out that we add a decimal to the quotient since we will be dividing tenths. Then have them divide the tenths. There are 3 tenths in each group, 9 tenths were used, and 2 tenths are left over. Show them how this is recorded; the number in each group, 3 tenths, goes above the line in the tenths place, and the 9 tenths (3 tenths x 3) go below in the correct places for ones and tenths. Subtracting 9 tenths from 11 tenths gives us the 2 tenths left over. The 2 tenths need to be replaced with 20 hundredths. On paper we just remember that it has to be combined with the 1 hundredths by writing the 1 hundredth next to the remainder. Now divide up the hundredths. There are 7 hundredths in each group, 21 hundredths were used, and no remainders. Show them where these numbers are recorded. Now we know that 1.11 ÷ 3 = 0.37.	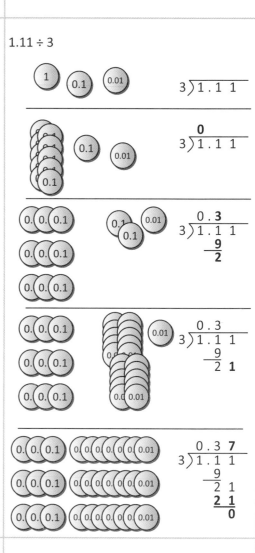
Write another problem where the quotient will be less than 1, such as **6.03 ÷ 9**. This time draw or use discs on the board, while the students record the steps on paper or the board. Do additional problems as needed. Exclude problems with remainders after the hundredths are divided for now.	6.03 ÷ 9 $$\begin{array}{r} 0.67 \\ 9\overline{)6.03} \\ 5\,4 \\ \hline 6\,3 \\ 6\,3 \\ \hline 0 \end{array}$$

Discussion	Text p. 56, Task 7
Rewrite the problem on the board and discuss the steps used in the division algorithm, as shown in the textbook. 7 tenths are divided by 2, with 3 tenths in each group, and the remaining tenth is renamed as 10 hundredths, so that there are 14 hundredths to divide by 2. Discuss some additional problems if needed, either selected from the next 3 tasks, or ones you make up. The quotient should be less than 1, with no remainders after hundredths are divided.	7. 0.37
Assessment	**Text p. 56, Task 8-10**

	8. (a) 0.13 (b) 0.21 (c) 0.11
	    (d) 0.17 (e) 0.15 (f) 0.16
	
	9. $0.15
	
	10. (a) $0.15 (b) $0.15 (c) $0.19
	     (d) $0.71 (e) $0.72 (f) $0.56

Practice	WB Exercise 35, pp. 77-78

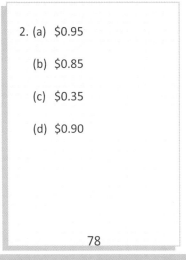

Exercise 35

1. (a) 0.24   (b) 0.21

  (c) 0.13   (d) 0.19

  (e) 0.28   (f) 0.17

  (g) 0.13   (h) 0.12

77

2. (a) $0.95

  (b) $0.85

  (c) $0.35

  (d) $0.90

78

2.3c Divide 2-Place Decimals

Objectives

♦ Divide 2-place decimals by a 1-digit whole number.

Note

In this lesson quotients may be greater than 1. Go through the division process using place-value discs and allow students to use the discs as needed. More capable students will not need to use physical discs and you can just talk them through the process. The steps are the same as those students learned for division of whole numbers. Rather than having any remainder ones, though, they can be renamed as tenths and further divided.

Illustrate the division algorithm	
Provide students with place-value discs and get them to divide them up and rename as needed. On the board relate what they are doing step-by-step with the division algorithm. Write the expression **7.02 ÷ 3** on the board and have students pick out 7 ones and 2 hundredths. Ask them to first divide the ones into 3 groups. There are 2 ones in each group; 6 are used up. Show them where to record this on the written problem, and how we find the remainder of 1. Ask students for the next step. The remainder one has to be renamed as tenths. There are no other tenths, so we record how many total tenths we have by writing the 0 next to the remainder of 1. Point out that it is not necessary to write the decimal point for the working under the line; we know it is tenths by aligning with the digit in the number we are dividing. However, it is imperative to write the decimal point above the line in the quotient if we have tenths to divide. Have students divide the tenths into the 3 groups now. There are 3 tenths in each group and 9 tenths are used. Continue to show how to record the steps. The remainder tenth will have to be renamed as one hundredth and the resulting 12 hundredths divided into the three groups. Now we know that 7.02 ÷ 3 = 2.34.	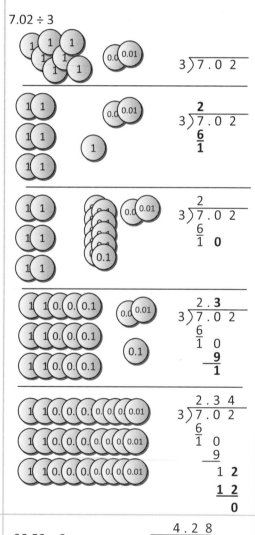
Write another problem where the quotient will be greater than 1, and no remainder hundredths, such as 38.52 ÷ 9. This time, draw or use discs on the board, while the students record the process. Do additional problems as needed.	38.52 ÷ 9 $$\begin{array}{r} 4.28 \\ 9{\overline{\smash{\big)}\,3\,8\,.\,5\,2}} \\ \underline{3\ 6} \\ 2\ \ 5 \\ \underline{1\ \ 8} \\ 7\ 2 \\ \underline{7\ 2} \\ 0 \end{array}$$

Discussion	Text p. 57, Task 11
Rewrite the problem on the board and discuss the steps used in the division algorithm, as shown in the textbook. Discuss some additional problems if needed, either selected from the next 3 tasks, or ones you make up. There should be no remainders after hundredths are divided.	11. 1.45
Assessment	**Text p. 57, Tasks 12-14**
	12.(a) 1.32 (b) 1.03 (c) 2.43 13. $2.30 14. (a) $1.55 (b) $2.75 (c) $3.45
Practice	WB Exercise 36, pp. 79-80

Exercise 36

1. (a) 4.13   (b) 3.22

   (c) 1.47   (d) 2.68

   (e) 22.75   (f) 5.27

   (g) 20.14   (h) 7.05

79

2. (a) $1.05   (b) $1.15

   (c) $1.45   (d) $1.35

   (e) $1.15   (f) $1.09

   (g) $2.55   (h) $1.75

80

2.3d   Append 0's to Divide

Objectives

♦ Divide decimal numbers by adding decimal places when there is a remainder.

Note

In this lesson students will be appending 0's to the dividend in order to continue division when there is a remainder. For slower students use place-value discs and discuss how the process, as done with the discs, is recorded in the division algorithm. More capable students will not need to use physical discs and you can just talk through the process. You can then possibly combine this lesson with some of the next lesson in order to be able to spend more time on the enrichment problems.

Illustrate the division algorithm

If needed provide students with place-value discs and get them to divide them up and rename as needed. On the board relate what they are doing step-by-step with the division algorithm.

Write the expression **112 ÷ 5** on the board. With student input go through the division process with the whole numbers. When you get to a remainder in the ones, remind them that in the past, when there was a remainder of ones, they wrote the answer as a quotient and a remainder, e.g., 112 ÷ 5 = 22 R 2. Now we don't have to stop with the remainder ones; we can rename them as tenths and continue to divide. We indicate this on the written problem by adding a decimal, and then a 0 for tenths. Instead of writing the quotient as a whole number with a remainder, we now can write it as a decimal.

$112 \div 5$

$$112 \div 5 = 22 \text{ R } 2$$

$$112 \div 5 = 22.4$$

Write the expression **66 ÷ 8** on the board. You can draw or use discs on the board while students record the steps. This time two 0's will be appended since both remaining ones and remaining tenths will be renamed.

$66 \div 8$

Discussion	Text p. 58, Task 15
Rewrite the problems on the board and discuss the steps used in the division algorithm, as shown in the textbook. Emphasize extending the number of decimal places by adding 0's.	15 (a) 1.25 (b) 1.35
Assessment	**Text p. 58, Tasks 16-17**

	16. (a) 6.08 (b) 1.5
	       6.08 1.5
	
	17. (a) 1.6 (b) 2.5 (c) 2.75
	    (d) 0.45 (e) 0.34 (f) 4.25

Practice	WB Exercise 37, pp. 81-82

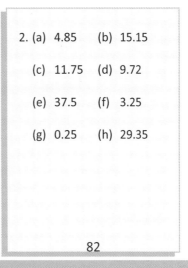

Exercise 36

1. (a) 1.4   (b) 0.75

  (c) 0.25   (d) 0.95

  (e) 1.24   (f) 1.25

  (g) 8.25   (h) 5.85

81

2. (a) 4.85   (b) 15.15

  (c) 11.75 (d) 9.72

  (e) 37.5   (f) 3.25

  (g) 0.25   (h) 29.35

82

2.3e   Estimate and Round Quotients

Objectives

♦ Estimate the quotient for division of decimals.
♦ Round the quotient to one decimal place.

Vocabulary

♦ Terminating decimal
♦ Non-terminating decimal

Note

Students may find it confusing that when we estimate for division, rather than rounding to a specific place value, we round to a multiple of the divisor. The reason we do so is to get a quotient with no remainder so that we can make a quick estimate.

It is not necessary for students to memorize the vocabulary terms *terminating* and *non-terminating* decimal.

Estimation	
Remind students that we can check the reasonableness of an answer by finding an estimate. For decimals in particular we can use an estimate to be sure that we placed the decimal point correctly.	
Ask students to estimate the answer to 428 ÷ 6. If they attempt to round the divisor to the nearest hundred or ten, let them go ahead and see if that makes it any easier to estimate the answer. Then remind them that in order to estimate the answer for division, we want to round to the nearest multiple of the number we are dividing by. We round 428 to 420, since 42 is a multiple of 6 (6 x 7 = 42). 420 is 42 tens, so the estimated answer is 7 tens, or 70.	428 ÷ 6 ↓ 420 ÷ 6 = 70 42 tens ÷ 6
Ask students to estimate the answer to 42.8 ÷ 6. This time we use 42 ÷ 6, and the estimated answer is 7. Then ask students to estimate the answer to 4.28 ÷ 6. We would not round 4.28 to 4, since we can't divide 4 by 6. We could round it to 4.2. Since this is 42 tenths, the estimated answer is 7 tenths.	42.8 ÷ 6 ↓ 42 ÷ 6 = 7 42 ones ÷ 6 4.28 ÷ 6 ↓ 4.2 ÷ 6 = 0.7 42 tenths ÷ 6
Point out that we can make a quicker estimate. Since 4.26 is less than 6, we know the answer will be less than 1. Since 4.26 is at least 1, we know the answer will have at least one tenth.	

Round the quotient	
Write the expression **428 ÷ 6** and discuss the division process. Since there is a remainder when we divide hundredths, we can continue the division to thousandths. Continue the division process until students realize that no matter how many decimal places we add, there will always be a remainder. The quotient is non-ending, or *non-terminating*. Tell students that in such cases we usually give an approximate answer by rounding the quotient. The quotient is 71.3 when rounded to 1 decimal place.	71.333 6)428.000 42 08 6 20 18 20 18 20 18 2
Write the expression **60 ÷ 7** and ask students to find the answer to one decimal place. Point out that to do so we have to first divide to two decimal places to know whether we round down or round up.	428 ÷ 6 ≈ 71.3

8.57
7)60.00
56
40
35
50
49
1

60 ÷ 7 ≈ 8.6

Discussion	Text p. 59, Tasks 18-19, 21
Tasks 18-19: Write the problems on the board and discuss finding the estimates. You can ask students to find the exact answer as well. For Task 19 point out that since 5.28 is slightly less than 6, we could also estimate the answer to simply be slightly less than 1 (6 ÷ 6). Task 21: Ask students to estimate the answer first. Since 7 is a little more than twice 3, the answer will be a little more than 2.	18. 4 (exact: 3.9) 19. 0.9, or 1 (exact: 0.88) 21. 2.3 2.3
Assessment	**Text p. 59, Tasks 20, 22-23**

	20. (a) 0.3 or between .1 and 1; 0.27
	    (b) 0.9 or almost 1 ; 0.89
	    (c) 5; 5.15
	
	22. 19.6
	
	23. (a) 0.2 (b) 0.6 (c) 0.6
	    (d) 0.2 (e) 0.4 (f) 5.4

Enrichment	
Ask students to carry out some of the divisions in Task 23 to more places to see if they get a terminating decimal or a non-terminating decimal and what repeating pattern the digits in the quotient have when the quotient is a non-terminating decimal.	22. (a) 0.1<u>6</u>6... (b) 0.<u>571428</u>5... (c) 0.<u>55</u>... (d) 0.225 (e) 0.41<u>66</u>... (f) 5.4375
Ask students to divide the numbers 1 through 8 by 9, as shown at the right. Then ask them what 9 ÷ 9 would be if we continued the pattern (0.9999...). Ask them if they think that 0.9999... with the 9's extended forever to smaller and smaller decimal places is the same as 1. (It is not necessary to provide a definitive answer at this time. Mathematically, 0.9999... is really 1. If you multiply both sides of $\frac{1}{3}$ = 0.3333... by 3, you get 1 = 0.9999...).	$1 ÷ 9 = 0.111...$ $2 ÷ 9 = 0.222...$ $3 ÷ 9 = 0.333...$ $4 ÷ 9 = 0.444...$ $5 ÷ 9 = 0.555...$ $6 ÷ 9 = 0.666...$ $7 ÷ 9 = 0.777...$ $8 ÷ 9 = 0.888...$ $9 ÷ 9 = 0.999...$?
Practice	WB Exercise 38, p. 83

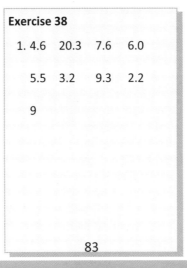

Exercise 38

1. 4.6 20.3 7.6 6.0

  5.5 3.2 9.3 2.2

  9

83

2.3f   Word Problems

Objectives

♦ Solve word problems involving division of decimals.

Note

In *Primary Mathematics* 3B the bar models for multiplication and division consisted of equal *units*. Use the idea of units here, and write the equations in terms of the value of multiple units and then the value of 1 unit.

Discussion	Text p. 60, Tasks 24-25
Write the problems on the board and discuss the solutions.	24. $3.20
Task 24: Ask students whether we are comparing two numbers: we are not. Ask them whether the problem asks for a part or a whole: we are given the whole. Then ask them whether we are given equal parts: there are 5 equal parts. So we can model the information by drawing a whole and dividing it into 5 equal parts, as shown in the textbook. Remind students that each equal part is called a *unit*. Write the equation **5 units = $8**.	25. $3.60 $3.60

Ask students what we want to find: the value of 2 units. Ask them what we need to find first: the value of 1 unit (this is labeled with a question mark in the diagram in the book). Continue the equations as shown at the right, asking students to supply the answers.	24. 5 units = $8
	     1 unit = $8 ÷ 5 = $1.60
	     2 units = $1.60 x 2 = $3.20

Tell students to suppose that the problem asked for the cost of 3 or 4 boxes instead of 2. Ask how we would find those answers. We would simply multiply the value for 1 unit by 3 or 4. Tell them that in many problems where we are given a whole and equal units, we can often solve them by first finding the value of 1 unit.	

Task 25: Ask students whether we are comparing two numbers. We are, since we are told how much money Taylor has compared to Bonita, so we can draw two bars to compare them. Ask if we are given equal units: we are. So we can draw 3 units for Taylor and 1 for Bonita, as shown in the textbook. Ask what we want to find: how much more money Taylor has than Bonita. Ask how many units that is: 2 units. Again, to answer the problem we first need to find the value of 1 unit. Then we can find the value of 2 units by subtraction, as the text does it, or by multiplying the value of 1 unit by 2.	25. 3 units = $5.40
	     1 unit = $5.40 ÷ 3 = $1.80
	     2 units = $1.80 x 2 = $3.60

Ask students to also find how much money they have altogether. Since there are 4 units total, we can find the total by multiplying the value for 1 unit by 4. We could also find the total units by adding how much money each has.	4 units = $1.80 x 4 = $7.20 They had $7.20 altogether.

Discussion	Text p. 61, Tasks 26-27

Task 26: In this task we have two parts, the amount of orange juice in the bottles and the amount left over. We are given the whole and the value of 1 part, the amount left over. Instead of just finding the value of the other part, though, we want to find the value of one of 5 equal units. To do so, we do have to find the value of the other part. So the model is a part-whole model, with one part divided into 5 equal units.	26. 0.95
	   0.95 gal
	
	27. 1.08
	   1.08 kg

	26. 5 units = 5 gal – 0.25 gal = 4.75 gal 1 unit = 4.75 gal ÷ 5 = 0.95 gal

Task 27: You can ask students for suggestions on how to model this task. Since all the flour in the 4 bags is used to make 5 cakes, we could draw two equal bars, one for the flour in the bags, which we divide into 4 equal parts, and one for the flour in each cake, which we divide into 5 equal parts.
 	
Ask students to explain why the solution in the book shows multiplication first. We need to find the total amount of flour first.
	27. 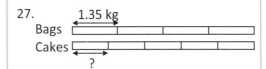
	   1 bag = 1.35 kg
	   4 bags = 1.35 kg x 4 = 5.4 kg
	   5 cakes = 4 bags = 5.4 kg
	   1 cake = 5.4 kg ÷ 5 = 1.08 kg

Enrichment	

More complex problems students will encounter later will involve making equal units between two bars divided up into different sized units.
 	
Discuss an alternate solution for Task 27 where the unequal units of the two bars are subdivided so that they now have equal units. If the unit in the bar with 4 equal units is divided into fifths, and that in the bar with 5 equal units is divided into fourths, there are now equal sized units in both bars. Each bag of flour is now represented by 5 units and each cake by 4 units. We know the value of 5 units and can find the value first of 1 unit and then of 4 units.
	27. (alternative solution)
	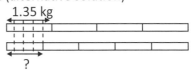
	   5 units = 1.35 kg
	   1 unit = 1.35 kg ÷ 5 = 0.27 kg
	   4 units = 0.27 kg x 4 = 1.08 kg
	   Or: 1.35 kg – 0.27 kg = 1.08 kg

Assessment	
Select some problems from Practice 2D on p. 62 of the textbook.	
Practice	WB Exercises 39-40, pp. 84-86

Exercise 39	Exercise 40	

1. 4 units = 1.48 m
   1 unit = 1.48 m ÷ 4 = **0.37 m**
   Each piece is 0.37 m long.
 	
2. 3 units = $20.40
   1 unit = $20.40 ÷ 3 = **$6.80**
   1 kg of shrimp costs $6.80.
 	
3. 5 units = $28.25
   1 unit = $28.25 ÷ 5 = **$5.65**
   Holly spent $5.65.
	1. $3.15 + $4.65 = $7.80
	   $7.80 ÷ 2 = **$3.90**
	   Each girl paid $3.90.
	
	2. 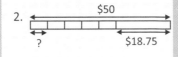
	   $50 – $18.75 = $31.25
	   $31.25 ÷ 5 = **$6.25**
	   1 kg of grapes cost $6.25.
		3. 2.7 lb – 1.2 lb = 1.5 lb
		   1.5 lb ÷ 5 = **0.3 lb**
		   Each block of butter weighs 0.3 lb.
		
		4. Total paint:
		   10.5 ℓ + 15.5 ℓ = 26 ℓ
		   Amount of paint in each can:
		   26 ℓ ÷ 4 = **6.5 ℓ**
		   Each can has 6.5 ℓ of paint.

84	85	86

Objectives

♦ Practice.

Note

Have students draw models only if needed. The need will vary with students.

Practice	Text p. 62, Practice 2D
	1. (a) 32.8 (b) 15.87 (c) 26.32

Text p. 62, Practice 2D

1. (a) 32.8 (b) 15.87 (c) 26.32
2. (a) $0.90 (b) $16.20 (c) $30.60
3. (a) 3.2 (b) 0.42 (c) 1.36
4. (a) $0.15 (b) $0.60 (c) $0.85
5. (a) 36; 39.24 (b) 30; 31.5 (c) 14; 13.58
6. (a) 3; 2.95 (b) 4; 3.99 (c) 9; 8.76

7. 4 bottles: 6 qt
   Amount in each bottle = 6 qt ÷ 4 = **1.5 qt**
   There were 1.5 qt in each bottle.

8. 1 liter: 1.25 kg
   6 liters: 1.25 kg x 6 = **7.5 kg**
   6 liters of gas weighs 7.5 kg.

9. 5 pieces: 6.75 yd
   1 piece: 6.75 yd ÷ 5 = **1.35 yd**
   Each piece is 1.35 yd long.

10. 1 pot hanger: $3 + $1.40 = $4.40
    4 pot hangers: $4.40 x 4 = **$17.60**
    It will cost him $17.60 to make 4 pot hangers.

11.

   6 units = 2.34 kg – 0.06 kg = 2.28 kg
   2.28 kg ÷ 6 = **0.38 kg**
   One bar weighs 0.38 kg.

12.

   3 units = $8.25
   1 unit = $8.25 ÷ 3 = $2.75
   2 units = $2.75 x 2 = **$5.50**
   or $8.25 – $2.75 = $5.50
   The book costs $5.50.

Enrichment

Write the problem at the right on the board and give students a chance to work on it on their own. There is more than one way to solve the problem. One method is suggested below. You can allow students to discuss their methods.	5 zucchini cost as much as 3 avocados. If 3 zucchini and 6 avocados cost $12.09, how much more does each avocado cost than each zucchini? (All avocados cost the same and all zucchini cost the same.)

We can start by simply diagramming what the problem states. Doing that can help us determine a method of solution.

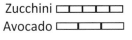

$12.09

In this case we can replace each set of 3 avocados with 5 zucchini. Then we can find the cost of one zucchini. Since 5 zucchini cost the same a 3 avocados, we can then find the cost of 1 avocado.

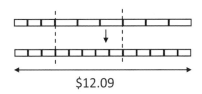

$12.09

13 zucchini = $12.09
1 zucchini = $12.09 ÷ 13 = $0.93
5 zucchini = $0.93 x 5 = $4.65
3 avocados = $4.65
1 avocado = $4.65 ÷ 3 = $1.55
$1.55 − $0.93 = $0.62

An avocado costs $0.62 more than a zucchini.

Objectives

♦ Practice.

Note

These problems are all word problems. You can save some to do at a different time for review as you go on with the next units. Have students draw models only if needed.

Practice	Text p. 63, Practice 2E

	1. 1 can: 5.5 ℓ
	   8 cans: 5.5 ℓ x 8 = **44 ℓ**
	   He used 44 ℓ of paint.
	
	2. Mrs. Bates = 4 daughters = 47.6 kg
	   1 daughter = 47.6 kg ÷ 4 = **11.9 kg**
	   Her daughter weighs 11.9 kg.
	
	3. Doll: $4.95
	   Robot = 3 dolls: $4.95 x 3 = **$14.85**
	   The robot costs $14.85.
	
	4. 3 girls paid: $17.40
	   1 girl paid: $17.40 ÷ 3 = **$5.80**
	   Each girl paid $5.80.
	
	5. 1 storybook: $2.80
	   5 storybooks: $2.80 x 5 = $14
	   Change: $20 − $14 = **$6**
	   He received $6 change.
	
	6.
	   5 units = $50 − $15.25 = $34.75
	   1 unit = $34.75 ÷ 5 = **$6.95**
	   1 m of cloth cost $6.95.
	
	7. Saved in 4 days: $4.60 x 4 = $18.40
	   Saved last day: $25 − $18.40 = **$6.60**
	   She saved $6.60 the last day.
	
	8. Cost of tea: $0.65 x 3 = $1.95
	   Cost of juice: $4.40 − $1.95 = **$2.45**
	   The glass of orange juice cost $2.45.

Enrichment - Mental Math Practice	Mental Math 20

Enrichment

Write the problem at the right on the board and give students a chance to work on it on their own. One method is suggested below. You can allow students to discuss their methods.	2 avocados and 6 zucchini together cost $8.60. 3 avocados and 3 zucchini together cost $7.80. How much more does each avocado cost than each zucchini? (All avocados cost the same and all zucchini cost the same.)

We can start by simply diagramming what the problem states. In the drawing at the right, avocados are the longer units. We do not yet know that avocados cost more than zucchini, but which gets the longer bars does not change the method of solution.

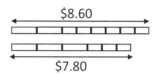

In this problem we cannot directly replace zucchini with avocado to get one type of unit. One bar is longer than the other, but subtracting one from the other does not give equal units. Rearranging the units does not help either. One way to approach a problem like this is to determine what information we can get out of the drawing.

In the first case there are an even number of each vegetable. So we can halve the number, and the total cost. Similarly, in the second case we have the same number of avocado and zucchini. So we can find the cost of 1 avocado and 1 zucchini.

2 avocados and 6 zucchini: $8.60
1 avocado and 3 zucchini: $8.60 ÷ 2 = $4.30
3 avocados and 3 zucchini: $7.80
1 avocado and 1 zucchini: $7.80 ÷ 3 = $2.60

Now that we have an equal number of avocados in both cases, the difference is now equal units of zucchini.

2 zucchini: $4.30 − $2.60 = $1.70
1 zucchini: $1.70 ÷ 2 = $0.85
1 avocado: $2.60 − $0.85 = $1.75
$1.75 − $0.85 = $0.90

An avocado costs $0.90 more than a zucchini.

Objectives

♦ Practice.

Note

These problems are all word problems and are fairly easy if students were successful with the problems in the previous exercise. By now, students should not need to draw careful models for simple problems of a type they have already seen; they can just draw simplified models to get them started if needed. Drawing the model is not the purpose of the problems, finding the answer is. If drawing models helps achieve the purpose, they should be used, but if the purpose can be achieved without them, then there is no reason to draw them just to draw them.

Practice	Text p. 64, Practice 2F
	1. 1 plate: $2.50 3 plates: $2.50 x 3 = **$7.50** They paid $7.50 altogether.
	2. 4 pillow cases: 6.6 m 1 pillow case: 6.6 m ÷ 4 = **1.65 m** She used 1.65 m of lace for each.
	3. 1.8 kg + (2.05 kg x 3) = 1.8 kg + 6.15 kg = **7.95 kg** The 4 packages weighed 7.95 kg.
	4. $5.65 + ($1.45 x 6) = $5.65 + $8.70 = **$14.35** She spent $14.35.
	5. 2 people: $6.70 1 person: $6.70 ÷ 2 = $3.35 $15.35 − $3.35 = **$12** Alice had $12 left.
	6. 3 units = $2.20 − $0.85 = $1.35 1 unit = $1.35 ÷ 3 = **$0.45** Each pencil cost $0.45.
	7. 1 unit = 3.46 m 3 units = 3.46 m x 3 = 10.38 m 10.38 m + 4.25 m = **14.63 m** She used 14.63 m of material.
	8. 4 kiwis: $2.20 1 kiwi: $2.20 ÷ 4 = $0.55 $0.60 − $0.55 = **$0.05** A kiwi is $0.05 cheaper at the sale.

Enrichment

Write the problem at the right on the board and give students a chance to work on it on their own. One method is suggested below. You can allow students to discuss their methods.	2 avocados and 5 zucchini together cost $7.95. 4 avocados and 3 zucchini together cost $10.65. How much more does each avocado cost than each zucchini? (All avocados cost the same and all zucchini cost the same.)
Start with a diagram.	
In this problem we cannot take a fraction of the total number and get whole amounts of avocados and zucchini, as in the enrichment problem in the previous lesson of this guide. But if we can get the same number of avocados or zucchini in both bars, then the difference will be equal units, as in the previous problem. See if students can come up with a way to manipulate one or both bars to get equal units of avocados or zucchini.	
One way to get equal units is to notice that the second bar has 4 avocados and the first has 2. So if we double the amount in the first case, we will have 4 avocados and 10 zucchini. The number of avocados is now the same for both cases. The units now can be lined up, and the difference between the bars is a whole number of units of zucchini.	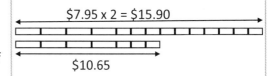
	7 zucchini: $15.90 − $10.65 = $5.25 1 zucchini: $5.25 ÷ 7 = $0.75 3 zucchini: $0.75 x 3 = $2.25 4 avocados: $10.65 − $2.25 = $8.40 1 avocado: $8.40 ÷ 4 = $2.10 $2.10 − $0.75 = $1.35
	An avocado costs $1.35 more than a zucchini.

2.3j   Review

Objectives

♦ Review.

Note

This review is cumulative. Questions 3, 5, 13-14, 17-18, and 25 review the just-completed unit. The rest of the questions review earlier material in *Primary Mathematics* 4A and 4B. Use the review to assess students' understanding and to determine if re-teaching is needed for any topics. Any solutions shown here are simply suggested approaches; students may solve a problem differently.

Review	Text pp. 65-67, Review B
1. (a) 124.66 (b) 124.57 (c) 124.46 (d) 124.55	14. (a) 6.7 (b) 14.3 (c) 4.9

Review

1. (a) 124.66
   (b) 124.57
   (c) 124.46
   (d) 124.55

2. (a) >   (b) <
   (c) <   (d) =

3. (a) 100   (b) 108
   (c) 10   (d) 1

4. (a) 100   (b) $35
   (b) 9 kg   (c) 10 m

5. (a) 29.4   (b) 62.3
   (c) 19.95   (d) 32.97

6. (a) 8000   (b) 80   (c) 8   (d) 80,000

7. 36

8. 8 x ? = 456 → 456 ÷ 8 = **57**
   The other number is 57.

9. (a) $\frac{4}{12} + \frac{9}{12} = \mathbf{1\frac{1}{12}}$ (b) $\frac{3}{9} + \frac{5}{9} = \mathbf{\frac{8}{9}}$

   (c) $\frac{5}{6} + \frac{3}{6} = 1\frac{2}{6} = \mathbf{1\frac{1}{3}}$ (d) $\frac{8}{10} + \frac{7}{10} = 1\frac{5}{10} = \mathbf{1\frac{1}{2}}$

10. (a) $\frac{11}{12} - \frac{3}{12} = \frac{8}{12} = \mathbf{\frac{2}{3}}$ (b) $\mathbf{2\frac{4}{7}}$

    (c) $\frac{4}{8} - \frac{1}{8} = \mathbf{\frac{3}{8}}$ (d) $\mathbf{5\frac{5}{6}}$

11. (a) $\frac{1}{12}, \frac{1}{3}, \frac{5}{6}$ (b) $1\frac{1}{4}, 1\frac{3}{4}, \frac{9}{4}$

    (c) $1\frac{3}{5}, 3, \frac{9}{2}$ (d) $2\frac{1}{5}, \frac{9}{4}, \frac{20}{6}$

12. $36,000

13. $77

Text pp. 65-67, Review B

14. (a) 6.7   (b) 14.3   (c) 4.9

15. Width: 50 cm^2 ÷ 10 cm = **5 cm**

16. Length of one side: 36 in. ÷ 4 = **9 in.**

17. $0.65 x 6 = **$3.90**
    He paid $3.90.

18. 4.8 m ÷ 8 = **0.6 m**.
    Each piece is 0.6 m long.

19. $\frac{7}{10}$ of the pole was not painted.

20. $\frac{3}{5}$ km x 4 = $\frac{3 \times 4}{5}$ km = $\frac{12}{5}$ km = $\mathbf{2\frac{2}{5}}$ **km**
    The perimeter of the garden is $2\frac{2}{5}$ km.

21. $\frac{3}{8} = \frac{6}{16}$; **6** out of 16 slices.

    Or: $\frac{3}{8}$ x 16 = 6

    She gave 6 slices away.

22. 8 units = 200
    1 unit = 200 ÷ 8 = 25
    3 units = 25 x 3 = **75**
    There were 75 girls.

23. $\frac{3}{4}$ of 6 ℓ = 6 x $\frac{3}{4}$ ℓ = $\frac{18}{4}$ = $\mathbf{4\frac{1}{2}}$ **ℓ**
    The bucket contains $4\frac{1}{2}$ ℓ of water.

24. $157 x 11 = $1727
    $1800 − $1727 = **$73**
    She saved $73 in the twelfth month.

25. $35.90 + $28.50 = $64.40
    $64.40 − $58.70 = **$5.70**
    He needs $5.70 more.

Enrichment

Write the problem at the right on the board and give students a chance to work on it on their own.	5 avocados and 4 zucchini together cost $18.00. 4 avocados and 5 zucchini together cost $17.10. How much more does each avocado cost than each zucchini? (All avocados cost the same and all zucchini cost the same.)
Start with a diagram. If students did the previous enrichment problems, they might immediately try to make equal numbers of avocados. This involves multiplying the first bar by 4 and the second by 5, so there are 20 avocados in both. After a lot of calculations they will find that the cost of 1 zucchini is $1.50 and the cost of 1 avocado is $2.40. If a student does not come up with the following alternate solution, discuss it with them. If we rearrange the units, we will find that the difference in the two bars is the same as the difference in one avocado and one zucchini, so only one calculation is needed to find the answer.	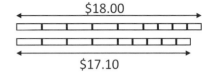 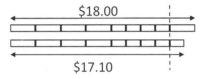 $18.00 − $17.10 = $0.90 An avocado costs $0.90 more than a zucchini.
Write the problem from the last lesson's enrichment again and see if students can find an alternate solution based on the idea above. The difference between the two bars, when the units have been rearranged, is the difference between 2 avocados and 2 zucchini. The difference between one avocado and one zucchini is half of that.	2 avocados and 5 zucchini together cost $7.95. 4 avocados and 3 zucchini together cost $10.65. How much more does each avocado cost than each zucchini? (All avocados cost the same and all zucchini cost the same.) 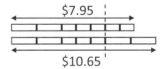 $10.65 − $7.95 = $2.70 $2.70 ÷ 2 = $1.35 An avocado costs $1.35 more than a zucchini.

3 Measures

Objectives

- Review conversion of measurements.
- Review expressing measurements in compound units.
- Review addition and subtraction of compound units.
- Multiply and divide compound units by 1-digit whole numbers.
- Solve word problems.

Suggested number of days: 7

		TB: Textbook WB: Workbook	Objectives	Material	Appendix
3.1	**Multiplication**				
3.1a	Review: Convert		◆ Review conversion of measurements. ◆ Review expressing measurements in compound units.		◆ Convert Measures (pp. a14-a16)
3.1b	Review: Add and Subtract		◆ Review addition of compound units. ◆ Review subtraction of compound units.		◆ Add and Subtract Measures (pp. a17-a19)
3.1c	Multiply	TB: pp. 68-69	◆ Multiply compound units by 1-digit whole numbers. ◆ Solve word problems.		◆ Multiply Measures (pp. a20-a21) ◆ Mental Math 21
3.2	**Division**				
3.2a	Divide	TB: pp. 70-71 WB: pp. 87-89	◆ Divide compound units by 1-digit whole numbers. ◆ Solve word problems.		
3.2b	Practice	TB: pp. 72-73	◆ Practice.		
3.2c	Review	TB: pp. 80-82 WB: pp. 90-93	◆ Review.		

Blank Page

3.1 Multiplication

Objectives

♦ Review conversion of measurements.
♦ Review expressing measurements in compound units.
♦ Review addition of compound units.
♦ Review subtraction of compound units.
♦ Multiply compound units by 1-digit whole numbers.
♦ Solve word problems.

Material

♦ Mental Math 21
♦ Appendix pp. a14-a21

Prerequisites

Students should be able to convert between measures within both the metric and U.S. customary systems and to add and subtract measures in compound units. They should also be familiar with mental math strategies used with 2-digit whole numbers, and mental math strategies for making 100 and making 1000. They should know the multiplication facts and how to multiply a multi-digit number by a 1-digit number.

Notes

In this part students will learn to multiply a measurement in compound units by multiplying each unit separately and then adding the products.

There are two distinct systems of measurement, the metric system and what is called the U.S. customary system in this curriculum. The metric system has become the global language of measurement and is used by about 95% of the world's population. It is used in science in the U.S. The metric system is based on powers of 10. If this is students' first exposure to the metric system, you should go back to *Primary Mathematics* 3B, including the Teacher's Guide, and do the units on measurement there in order to become familiar with the metric system. There is not adequate review in this part to really understand or get a "feel" for the metric system of measurement.

In *Primary Mathematics* 3B students learned to add and subtract measurements in compound units. Measurements in compound units are those that involve two units, such giving the length of an object in meters and centimeters, or in feet and inches. Adding and subtracting measures in compound units allows students to both practice measurement unit conversions and mental math. The first two lessons in this part therefore review conversion of measurement and adding and subtracting measures in compound units.

In the metric system it is easy to convert between units, so two measurements in, for example, meters and centimeters can be easily converted to centimeters only and then added together. However, adding and subtracting by making the next larger unit or taking from the next larger unit is very similar to using mental math strategies. So measures in compound units can be added or subtracted by first adding the values for the larger units, and then the values for the smaller units using mental math strategies for making 100 or 1000. Students less adept at mental math can simply convert or "rename" one of the larger units.

4 m 25 cm + 5 m 90 cm: add the meters: 9 m; add the cm: 1 m 15 cm; answer is 10 m 15 cm.

In the U.S. customary system it is not as easy to convert between units and the conversion factors vary. One foot is 12 inches, but 1 yard is 3 feet. So it makes even more sense to first add the larger units together, and then the smaller ones. Then, we only have to convert one of the larger units. Or, we can apply mental math strategies similar to those learned for base-10. This will give students some experience working with other bases than base-10, such as "making a 3" or "making a 12."

3 ft 10 in. + 5 ft 7 in.: add the ft: 8 ft; add the inches: 1 ft 5 in.; answer is 9 ft 5 in.

Students have learned that in order to multiply a multi-digit number, such as 485, by a 1-digit number, such as 8, the digit in each place value is multiplied by 8 and then the products are added together, renaming as needed.

485 x 8 = (400 x 8) + (80 x 8) + (5 x 8) = 3200 + 640 + 40 = 3880

(This process is simplified by using the standard algorithm for multiplication.)

Similarly, measurements in compound units can be multiplied by a 1-digit number by multiplying the value for each unit of measurement separately and then adding the products together, converting as needed.

4 kg 850 g x 8 = (4 kg x 8) + (850 g x 8) = 32 kg + 6800 g = 32 kg + 6 kg + 800 g = 38 kg 800 g

4 ft 8 in. x 8 = (4 ft x 8) + (8 in. x 8) = 32 ft + 64 in. = 32 ft + 5 ft 4 in. = 37 ft 4 in.

With the metric system it is just as easy to convert to the smaller unit and multiply (using the standard multiplication algorithm or mental math strategies) and then convert back.

4 kg 850 g x 8 = 4850 g x 8 = 38,800 g = 38 kg 800 g

We could do this also with the U.S. customary system, but that often results in more computation steps or dividing a larger number to convert back.

4 ft 8 in. x 8 = 56 in. x 8 = 448 in. = 37 ft 4 in.

(4 ft x 12 + 8 in. = 48 in. + 8 in. = 56 in.)   (448 ÷ 12 = 37 R 4)

Doing operations such as addition, subtraction, multiplication, and division on compound measures by working with each unit of measurement separately and then "renaming" as needed is a good way to practice measurement conversions as well as mental math strategies, and also a good concrete introduction to working with other bases than 10, such as base-60 (for time) and the bases 3, 4, 12, and 16 for yards to feet, gallons to quarts, feet to inches, and pounds to ounces.

3.1a Review: Convert

Objectives

♦ Review conversion of measurements.
♦ Review expressing measurements in compound units.

Vocabulary

♦ Conversion factor
♦ Centi-
♦ Milli-
♦ Kilo-

Note

This lesson reviews conversion of measures and compound units. If students are already comfortable with this topic, you can skip this lesson or combine it with the next one, which is also review.

For more extensive review, use *Primary Mathematics* 3B.

Review conversion factors	
Ask students if they remember units that are used to measure length, weight, or capacity. When they name one, point to an object or name an object and ask them to estimate its measurement in that unit. For example, if they say feet, ask them to estimate the length of the classroom in feet. If they say kilograms, ask them to estimate their own weight in kilograms.	1 km = ___ m 1 m = ___ cm 1 kg = ___ g 1 ℓ = ___ ml 1 yd = ___ ft 1 ft = ___ in. 1 lb = ___ oz 1 gal = ___ qt 1 qt = ___ pt 1 pt = ___ c
Remind students that they are learning about two distinct measurement systems, the metric system, which is used in most of the world, and the U.S. customary system. Let them talk about which system they might prefer and why. We also measure time in units, and the same system is used everywhere.	1 year = ___ months 1 week = ___ days 1 day = ___ hours 1 h = ___ minutes 1 min = ___ seconds
Write the problems at the right on the board and ask students to supply the conversion factors. You can review the abbreviations at this time as well. Measurements are usually no longer abbreviated with periods even for the U.S. customary system, except for the abbreviation for inches, in., to distinguish it from the word in. Abbreviations are not standardized. Though this curriculum uses a cursive ℓ for liters others use an uppercase L and some a lower case non-cursive l. Different abbreviations may be used for time, such as "s" or "sec" for seconds and "h" or "hr" for hours.	1 km = 1000 m 1 m = 100 cm 1 kg = 1000 g 1 ℓ = 1000 ml
Point out that the conversions in the table at the top of the page go from the larger unit to the smaller. 1 meter is the larger unit, and 1 centimeter is the smaller unit. 100 is called the *conversion factor* for meters to centimeters.	1 yd = 3 ft 1 ft = 12 in. 1 lb = 16 oz 1 gal = 4 qt 1 qt = 2 pt 1 pt = 2 c
Point out that that for the metric system, the conversion factor meters to centimeters is 100, but for the other common ones listed here it is 1000. If students know what the prefixes mean, that will help them remember what conversion factor to use. Centi - means 100; 1 meter is 100 *centi*meters. Kilo- means 1000, and is used for the larger measurement. So 1 *kilo*meter is 1000 meters, and 1 *kilo*gram is 1000 grams. Milli- also means 1000 but is used for the smaller measurement, so 1 liter = 1000 *milli*liters.	1 year = 12 months 1 week = 7 days 1 day = 24 hours 1 h = 60 minutes 1 min = 60 seconds

Review conversion of measurements	**Appendix p. a14**

Provide students with appendix p. a14. Discuss the problems.

Problems 1-4: When converting from a larger unit to a smaller unit of measurement, we are essentially cutting up the unit into smaller ones. We use the conversion factor to find out how many smaller units we have. Since 1 ft is 12 inches, then 4 ft is 4 times 12 inches. This is similar to using the part-whole model for multiplication in word problems where we have the value of 1 unit and need to find the value of more units. You can illustrate a few problems this way, but do not ask students to draw models for all conversion problems.

The similarity in the numerical portion of the answers to 3 and 4 could lead to a discussion on how to solve these problems mentally using factors or number bonds. For example, 8 x 24 is (8 x 20) + (8 x 4), or 24 x 2 x 2 x 2.

Problems 5-8: Remind students that measurements can be made in *compound units*. For example, if the capacity of a jug is more than 4 liters but less than 5 liters, the amount between 4 and 5 liters can be measured in milliliters. In these problems we want to convert the compound measurement to the measurement in just the smaller unit. To do so with 4 ℓ 250 ml all we need to convert is the part that is 4 ℓ, and then add that to 250 ml.

Problem 9: In this problem we are converting from a smaller unit to a larger unit. Ask students to find the number of feet in 48 inches. Make sure they understand that we are grouping the inches by 12's, since 1 group of 12 inches is 1 ft. So we divide. 48 inches divided into groups of 12 gives 4 groups, so there are 4 ft in 48 inches. Then ask them to find the number of feet in 50 inches. Now when we divide by 12, there will be 2 inches left over. We can think of this as splitting 50 into the number of inches that would make a whole number of feet, and the rest. There are not an exact number of feet in 50 inches and we express the answer as a compound unit; feet and the left-over inches.

Problems 10-14: Point out that in going from a smaller unit, many units, to a larger unit, fewer units, the number of units in the answer is smaller. It is the number of groups we can make, so we divide by how many go in each group.

1. 1 ft = **12** in.
   4 ft = 4 x **12** in. = **48** in.

   12 in. 1 unit (1 ft) = 12 in.
   4 units (4 ft) = 4 x 12 in.= 48 in.

2. 1 m = **100** cm
   9 m = 9 x **100** cm = **900** cm
3. 1 day = **24** hours
   8 days = 8 x **24** hours = **192** hours
4. 1 lb = **16** oz
   12 lb = 12 x **16** oz = **192** oz

5. 4 ℓ 250 ml
   = (4 x **1000**) ml + 250 ml
   = **4250** ml
6. 5 km 40 m
   = (5 x **1000**) m + **40** m
   = **5040** m
7. 4 years 5 months
   = (4 x **12**) months + 5 months
   = **48** months + 5 months
   = **53** months
8. 4 hours 20 minutes
   = (4 x **60**) minutes + 20 minutes
   = **240** minutes + 20 minutes
   = **260** minutes

9. **12** in. = 1 ft
   48 in. = 48 ÷ 12 = **4** ft
   50 in. = **4** ft **2** in.   (50 ÷ 12 = **4** R **2**)

10. **3** ft = 1 yd
    8 ft = **2** yd **2** ft   (8 ÷ 3 = 2 R 2)
    /\
    6 2
11. **100** cm = 1 m
    602 cm = **6** m **2** cm
    /\
    600 2
12. **1000** g = 1 kg
    2400 g = **2** kg **400** g
    /\
    2000 400
13. **4** qt = 1 gal   365 ÷ 4 = **91** R **1**
    365 qt = **91** gal **1** qt
14. **24** h = 1 day   365 ÷ 24 = **15** R **5**
    365 h = **15** days **5** h

Practice	**Appendix p. a15**

Answers are on appendix p. a16.

3.1b   Review: Add and Subtract

Objectives

♦ Review addition of compound units.
♦ Review subtraction of compound units.

Note

This lesson reviews addition and subtraction of compound units. This is a good way to practice conversions if you skipped the previous lesson.

You can do the problems in this lesson on the board so that slower students can follow the discussion. Or, for more capable students, you can provide them with copies of appendix p. a17 and only discuss the first few problems in each set and have students do the rest independently for assessment.

Review subtraction of compound units	Appendix p. a17
Discuss the problems at the right. To subtract the smaller units from the larger one, we need to convert the larger measurement to the smaller. Point out that we can use mental math strategies for "making 100" or "making 1000" with the metric system measurements. We can "make" other numbers with the other measurements, like "make 12" when subtracting inches from feet.	1 m – 42 cm = ____ cm **(58 cm)** 1 km – 390 m = ____ m **(610 m)** 1 ℓ – 7 ml = ____ ml **(993 ml)** 1 ft – 7 in. = ____ in. **(5 in.)** 1 yd – 2 ft = ____ ft **(1 ft)** 1 lb – 7 oz = ____ oz **(9 oz)** 1 h – 15 min = ____ min **(45 min)**
When we have more than 1 unit to subtract from, rather than convert all of the larger measurements into the smaller ones, we just need to convert one of them. With the U.S. customary units it would be a pain to convert 19 ft to 228 in., subtract 3 in., and then convert back using division. It is much easier to split 19 ft into 18 ft and 1 ft and subtract 3 in. from the 1 ft.	4 m – 25 cm = ____ m ____ cm **(3 m 75 cm)** 10 km – 90 m = ____ km ____ m **(9 km 910 m)** 3 ℓ – 985 ml = ____ ℓ ____ ml **(2 ℓ 15 ml)** 19 ft – 3 in. = ____ ft ____ in. **(18 ft 9 in.)** 8 lb – 10 oz = ____ lb ____ oz **(7 lb 6 oz)** 5 gal – 3 qt = ____ gal ____ qt **(4 gal 1 qt)**
To subtract from a compound unit, we can subtract from the larger unit, then add back in the smaller unit, or "rename" the larger unit; both strategies students have learned with whole numbers.	4 h 10 min – 25 min = ___ h ___ min **(3 h 45 min)** (4 h – 25 min + 10 min; subtract 25 min from 4 h 3 h 35 min + 10 min) 19 ft 1 in. – 3 in. = ____ ft ____ in. **(18 ft 10 in.)** (19 ft – 3 in. + 1 in.; subtract 3 in. from 19 ft 18 ft 9 in. + 1 in.)
To subtract a compound unit from a compound unit, we can first subtract the larger unit, and then the smaller unit.	6 h 10 min – 2 h 25 min = ___ h ___ min **(3 h 45 min)** (6 h 10 min – 2 h = 4 h 10 min 4 h 10 min – 25 min = 3 h 45 min) 12 m 45 cm – 7 m 80 cm = ___ m ___ cm **(4 m 65 cm)** (12 m 45 cm – 7 m = 5 m 45 cm 5 m 45 cm – 80 cm = 4 m 20 cm + 45 cm = 4 m 65 cm)

Review addition of compound units	**Appendix p. a17**
Continue discussion of the problems at the right.	65 cm + 40 cm = _____ m _____ cm (**1** m **5** cm)
We can simply add and then convert. Or, we can add mentally by "making" the conversion factor. This is shown with the number bonds. In the example at the right for adding meters, it might be easier to simply add, rather than trying to make 1000. If a student is good at mental math, it should be easy to make 12, 3, 4, or 16 with the U.S. customary measurements. Allow students to use either method, but encourage them to use mental math strategies when feasible.	/\ 5 60 780 m + 390 m = _____ km _____ m (**1** km **170** m) /\ 220 170 400 ml + 750 ml = _____ ℓ _____ ml (**1** ℓ **150** ml) /\ 600 150
When both units are the same, or only one addend has a compound unit, we can just add the smaller units together. Sometimes, that makes one more of the larger unit, in which case we "rename" and add to the larger unit, if there is one.	9 in. + 6 in. = _____ ft _____ in. (**1** ft **3** in.) /\ 3 3 8 oz + 10 oz = _____ lb _____ oz (**1** lb **2** oz) /\ 2 6 3 ft 7 in. + 10 in. = _____ ft _____ in. (**4** ft **5** in.) /\ 5 2 8 h 30 min + 20 min = ___ h ___ min (**8** h **50** min)

To add a compound unit to a compound unit, first add the larger unit, and then the smaller. This is much easier than converting both measurements completely to the smaller unit, particularly with U.S. customary units.	3 ft 7 in. + 6 ft 10 in. = ___ ft ___ in. (**10** ft **5** in.)
	 (3 ft 7 in. + 6 ft = 9 ft 7 in.
	    9 ft 7 in. + 10 in. = 10 ft 5 in.)
	
	 7 km 900 m + 7 km 105 m = __ km __ m (**15** km **5** m)
	 (7 km 900 m + 7 km = 14 km 900 m
	    14 km 900 m + 105 m = 15 km 5 m)

Practice	**Appendix p. a18**
Answers are on appendix p. a19.	

3.1c Multiply

Objectives

♦ Multiply compound units by 1-digit whole numbers.
♦ Solve word problems.

Note

Although the solutions for Tasks 1 and 2 are written out in detail, most students should be able to find the answers mentally. Do not require students to write each little step when working on their own; encourage them to use mental math. For slower students go through each step but also encourage them to combine steps when they understand the process and are capable of doing so.

Discussion	Text p. 68
Remind students that multiplication can be thought of as repeated addition. We could add the weight of the three packages on p. 68 together. To do this, we would add the kilograms together, add the grams together, and then add both sums together. Essentially we are multiplying the kilogram by 3, the 200 grams by 3, and adding the products together. So we can multiply each part of the compound measurement separately. Tell students that we now have 6 packages of the same weight and ask them for the total weight. This time, when we multiply the value for the smaller measurement unit, grams, by 6, we get more than 1000 g, or 1 kg. So before we add it to the product of the kilograms and 6, we need to convert it to kilometers and meters.	**3** kg **600** g **3** kg **600** g 1 kg 200 g　　　　1 kg 200 g 1 kg 200 g　　　　／　＼ + 1 kg 200 g　　　1 kg　　200 g 3 kg 600 g　　　x 3　　　x 3 　　　　　　　　3 kg　　600 g 　　　　　　　　　＼　／ 　　　　　　　　　3 kg 600 g 1 kg　200 g　　　1 kg 200 g 1 kg　200 g　　　／　＼ 1 kg　200 g　　　1 kg　　200 g 1 kg　200 g　　　x 6　　　x 6 1 kg　200 g　　　6 kg　1200 g + 1 kg　200 g　　　＼　／ 6 kg 1200 g　　　7 kg 200 g 7 kg　200 g

Word Problems	Text p. 69, Tasks 1-2
Discuss these two problems with students. After multiplying the units separately, we need to carry out any necessary conversions. The answers at the right show the steps in solving the problems, but you should allow students to find the answer mentally, if they can.	1. 1 km 300 m x 4 = (1 km x 4) + (300 m x 4) = 4 km + 1200 m = 4 km + 1 km 200 m = **5** km **200** m **5** km **200** m 2. 2 ℓ 400 ml x 5 = (2 ℓ x 5) + (400 ml x 5) = **10** ℓ **2000** ml = 10 ℓ + 2 ℓ = **12** ℓ **12** ℓ

Assessment	
Write the following expressions at the right on the board and have students solve them. You can have them discuss their solutions and see if they came up with any short-cuts.	6 h 30 min x 6 (39 h)
	Since 30 min is half an hour, 30 min x 2 = 1 h, so 30 min x 6 = 3 h. 36 h + 3 h = 39 h. Or: 30 x 6 = 3 x 10 x 6 = 3 x 60 = 3 h, 36 h + 3 h = 39 h
	2 ft 10 in. x 5 (14 ft 2 in.)
	3 lb 8 oz x 4 (14 lb)
	Since 8 oz is half of a pound, 4 of them are 2 lb. 12 lb and 2 lb is 14 lb.
	16 gal 3 qt x 6 (100 gal 2 qt)
	4 kg 201 g x 5 (21 kg 5 g)
Enrichment - Mental Math Practice	Mental Math 21
Practice	**Appendix p. a20**
Answers are on appendix p. a21.	

3.2 Division

Objectives

♦ Divide compound measures by 1-digit whole numbers.
♦ Solve word problems.

Prerequisites

Students should be able to add, subtract, and multiply measures in compound units and convert between measures within both the metric and U.S. customary systems. They should also be familiar with mental math strategies used with 2-digit whole numbers. They should know division facts and how to divide a multi-digit number by a 1-digit number.

Notes

In this part students will learn how to divide measures in compound units.

To divide measures in compound units, we can split the measure into the larger and smaller unit and divide each one separately. Since division of the larger unit will sometimes result in a remainder, we need to "rename" or convert the remainder and combine with the smaller unit, and then divide. This process will allow students to both review division and see how the concept behind the division algorithm (divide the digit in the largest place value first, rename the remainder, and add to the digit in the next place value and divide that, and so on) applies to situations other than division of whole numbers. At the secondary level they will use the same algorithm to divide polynomial expressions.

For example, to divide 7 hours and 30 minutes by 6, we first divide the hours by 6, get a remainder of 1 hour, rename it to minutes, add this remainder to the 30 minutes, and then divide the minutes by 6. This process is similar to the division algorithm, but since the number system is base-60, not base-10 as with whole numbers, we can't simply "drop" the 3 down next to the remainder of 1 hour. We have to first convert the hour to 60 minutes and add that to the 30 minutes.

Blank Page

3.2a   Divide

Objectives

♦ Divide compound measures by 1-digit whole numbers.
♦ Solve word problems.

Note

In all of these problems there will be no remainder after the smaller unit is divided. If an observant student asks what to do if there is a remainder, tell them we can express the answer as a decimal, or in some cases convert to an even smaller unit.

Most students should be able to follow the connection between the process of dividing compound measures and the division algorithm. For struggling students you may want to omit the enrichment section.

Discuss estimation	Text p. 70
Tell students that we can also divide measures in compound units by dividing each unit separately, but with division we have to deal with remainders when a number does not divide evenly. In the example in the text we first divide 5 meters by 4. This leaves a remainder of 1 meter. So we have 1 meter and 20 centimeters that still need to be divided. Ask students how we can divide the remaining amount. We can convert the remainder of 1 m to 100 centimeters and combine it with the 20 centimeters. We can then divide 120 centimeters by 4. Tell students that instead of the ribbon being 5 m 20 cm long, it is 25 ft 8 in. long and ask them to divide by 4. We first divide 25 ft by 4, which gives 6 ft with 1 ft left over. To divide the left-over foot by 4, we need to change it into inches. We can add that to the inches we already have, and divide the inches by 4. Or we could divide the 12 inches by 4, the 8 inches by 4, and add the two quotients.	**1 m 30 cm** **1 m 30 cm** 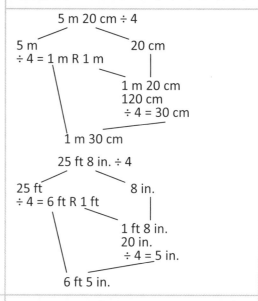

Word problems	Text p. 71, Tasks 1-3

Guide students through the steps for solving these division problems.
 	
Task 1: The larger unit can be evenly divided by 5, so we just divide the smaller unit as well by 5.
 	
Task 2: Dividing 7 h by 6 gives 1 h with a remainder of 1 h, or 60 min. That is added to the 30 min and then the total minutes are divided by 6.
 	
Task 3: In this case we cannot divide 3 liters by 8 and get a whole liter, so all 3 liters are converted to milliliters before the total milliliters are divided.
	1. 5 kg 650 g ÷ 5 = ?
	   5 kg ÷ 5 = 1 kg 650 g ÷ 5 = 130 g
	   1 kg 130 g
	   1 kg 130 g
	
	2. 7 h 30 min ÷ 6 = ?
	   7 h ÷ 6 = 1 h remainder 1 h (60 min)
	               60 min + 30 min = 90 min
	                           90 min ÷ 6 = 15 min
	   1 h 15 min
	   1 h 15 min
	
	3. 3 ℓ 200 ml ÷ 8 = ?
	           3 ℓ 200 ml = 3200 ml
	           3200 ml ÷ 8 = 400 ml
	   400 ml
	   400 ml

Enrichment

You can show students how dividing compound measures is similar to division of whole numbers. Refer back to the problem on p. 70. To divide 520 by 4, we first divide the hundreds and get a remainder of 1 hundred which is renamed as 10 tens and added to the 2 tens. Then 12 tens are divided by 4. Similarly, the remainder from the division of the larger unit is "renamed" and added to the smaller unit. Except that the remainder is renamed as 100 cm.

Now refer to Task 2. We can divide as with the division algorithm, but the remainder of 1 hour is renamed as 60 minutes and then added to the 30 minutes.

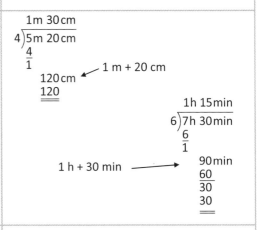

Assessment

Write the following problems on the board and ask students to solve them. The division algorithm representation is shown here and will sometimes be shown in the solutions in this guide since it is a concise way to show how the problem can be solved.

⇒ 10 min 12 s ÷ 6   (1 min 42 s)
⇒ 20 lb 2 oz ÷ 7   (2 lb 14 oz)

⇒ A piece of rope 22 ft long needs to be cut into 3 pieces. The second piece has to be twice as long as the first piece and the third piece has to be three times as long as the first piece. How long will the second piece be?
              (The second piece will be 7 ft 4 in. long.)

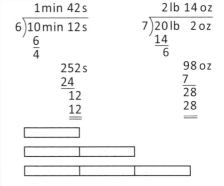

6 units = 22 ft
1 unit = 22 ft ÷ 6 = 3 ft R 4 ft
    4 ft = 48 in.   48 in. ÷ 6 = 8 in.
1 unit = 3 ft 8 in.
2 units = 3 ft 8 in. x 2 = 7 ft 4 in.
Or: The second rope is 2 out of 6 units, or one third of the total.
2 units = 22 ft ÷ 3 = 7 ft R 1 ft
              12 in. ÷ 3 = 4 in.
2 units = 7 ft 4 in.

Practice

WB Exercise 41, pp. 87-89

Exercise 41

1. (a) 4 ℓ 900 ml
   (b) 7 m 95 cm
   (c) 31 km 250 m
   (d) 1 kg 100 g
   (e) 1 h 50 min
   (f) 400 ml
   (g) 25 lb 2 oz
   (h) 11 ft 4 in.

2. 1 ℓ 500 ml x 3
   = 3 ℓ 1500 ml = **4 ℓ 500 ml**
   The bucket holds 4 ℓ 500 ml.

87

3. 5 kg 500 g x 6
   = 30 kg 3000 g = **33 kg**
   The total weight is 33 kg.

4. 1 h 40 min x 4
   = 4 h 160 min = **6 h 40 min**
   It takes 6 h 40 min to wash 4 loads.

5. 6 kg 750 g ÷ 9
   6750 g ÷ 9 = **750 g**
   Each packet weighed 750 g.

88

6. (a) 4 m 50 cm ÷ 3
       = 3 m 150 cm ÷ 3
       = **1 m 50 cm**
       Each piece of wire was 1 m 50 cm long.
   (b) 1 m 50 cm x 2
       = 2 m 100 cm = **3 m**
       He used 3 m of wire.

7. 6 kg 850 g – 600 g = 6 kg 250 g
   6 kg 250 g ÷ 5
   = 5 kg 1250 g ÷ 5 = **1 kg 250 g**
   Each book weighs 1 kg 250 g.

89

Objectives

♦ Practice.

Note

By now students should be able to solve the simpler ones without a model and just use models for more complex word problems. Rather than forcing a student to draw a model when he or she can solve the problem easily without one, provide the student with some more challenging problems from one of the supplementary books where a model is needed more. If a student struggles, though, encourage the student to draw a model for the problem.

Practice	Text pp. 72-73, Practice 3A
	1. (a) 3 km 200 m x 5 = 15 km 1000 m = **16 km** (b) 4 ℓ 300 ml x 4 = 16 ℓ 1200 ml = **17 ℓ 200 ml** (c) 2 h 20 min x 5 = 10 h 100 min = **11 h 40 min** (d) 5 kg 200 g x 3 = 15 kg 600 g = **15 kg 600 g** (e) 6 m 20 cm x 6 = 36 m 120 cm = **37 m 20 cm** (f) 3 yd 2 ft x 7 = 21 yd 14 ft = **25 yd 2 ft** 2. (a) 2 ℓ 240 ml ÷ 2 = **1 ℓ 120 ml** (b) 5 km 300 m ÷ 2 = 4 km 1300 m ÷ 2 = **2 km 650 m** (c) 1 h 30 min ÷ 5 = 90 min ÷ 5 = **18 min** (d) 4 kg 500 g ÷ 3 = 3 kg 1500 g ÷ 3 = **1 kg 500 g** (e) 2 m 60 cm ÷ 4 = 260 cm ÷ 4 = **65 cm** (f) 4 ft 3 in ÷ 3 = 3 ft 15 in. ÷ 3 = **1 ft 5 in.** 3. 1 ℓ 275 ml x 2 = **2 ℓ 550 ml** She used 2 ℓ 550 ml of syrup. 4. 3 kg 570 g ÷ 3 = **1 kg 190 g** The beans in each bag weighed 1 kg 190 g. 5. 3 h 30 min x 5 = 15 h 150 min = **17 h 30 min** He spent 17 h 30 min painting his house. 6. (a) 1 kg 800 g x 3 = 3 kg 2400 g = **5 kg 400 g** The watermelon weighs 5 kg 400 g. (b) 5 kg 400 g + 1 kg 800 g = **7 kg 200 g** The total weight is 7 kg 200 g. 7. (a) 8 h 30 min x 6 = 48 h 180 min = **51 h** In 6 days she works 51 hours. (b) 51 x $5 = **$255** She earns $255. 8. 3 units = 3 m 66 cm 1 unit = 3 m 66 cm ÷ 3 = 1 m 22 cm 2 units = 1 m 22 cm x 2 = **2 m 44 cm** The longer piece was 2 m 44 cm long.

9. Total sugar: 1 kg 240 g + 1 kg 160 g = 2 kg 400 g

   2 kg 400 g ÷ 8 = 2400 g ÷ 8 = **300 g**

   She used 300 g of sugar for each cake.

10. Total ribbon: 3 m 50 cm x 2 = 6 m 100 cm = 7 m

    7 x $4 = **$28**

    She paid $28 for the ribbon.

Enrichment

List the conversions for time. Point out that the list is in order from the largest unit of measurement, year, to the smallest one, second. Ask students how they could use the list to determine conversion factors for other times, such as finding how many seconds there are in a day.

Since there are 24 hours in a day, and 60 minutes in an hour, then there are 60 x 24 minutes in a day. Similarly, there are 60 x 60 x 24 seconds, or 86,400 seconds, in a day. We multiply the conversion factors for days to hours, hours to minutes, and minutes to seconds together to get one for days to seconds.

Ask students to find the number of seconds in a year. We can't go from months to weeks so this chart does not help. But we can use the fact that there are 365 days in a non leap-year. The number of seconds in a year is 365 x 24 x 60 x 60 = 31,536,000 seconds (about thirty-one and a half million seconds).

We can use this same concept to find the number of cups in a gallon or quart. If we add in a half-gallon, then we can just multiply by 2's for all conversions with U.S. customary units for capacity.

Students may be interested in other units of measurement in the metric system. They may already know that there are 1000 millimeters in a meter. They can research other units between millimeter and kilometer, such as hectometer, decameter, and decimeter, or smaller units of measurements, such as picometers, nanometers, or angstroms. There are also units of measurement for many other things than length, weight, and capacity, such as paper sheet sizes, steel metal gauges, tennis racket gauges, hurricane intensity, solar flare intensity, and others.

1 year = 12 months
1 week = 7 days
1 day = 24 hours
1 hour = 60 minutes
1 minute = 60 seconds

1 day = 24 x 60 x 60 seconds
       = 86,400 seconds

1 year = 365 days (non leap-year)

1 year = 365 x 86,400 seconds
        = 31,536,000 seconds

1 gallon = 2 half-gallons
1 half-gallon = 2 quarts
1 quart = 2 pints
1 pint = 2 cups

1 gallon = 2 x 2 x 2 x 2 cups
          = 16 cups

Objectives

♦ Review.

Note

In the workbook there are two reviews after the last exercise for Unit 3, but in the textbook there are two reviews after Unit 4. You can do one of the textbook reviews and one of the workbook reviews now and the other after Unit 4.

Review	Text pp. 80-82, Review C
1. (a) 2 (b) 0.03 or $\frac{3}{100}$ 2. (a) 90,504 (b) 17,541 3. 2.69 4. (a) 80,300; 82,300 (b) 5.59, 6.09 5. (a) 14,680 (b) 30,083 (c) 9900 (d) 89,301 6. (a) 7.03 (b) 4.9 (c) 2.41 (d) 3.602 7. (a) 14,058; 14,508; 41,058; 41,508 (b) 0.96, 8.54, 24.3, 72 8. (a) 0.28 (b) 0.04 9. (a) A: 4490 B: 4540 C: 4620 (b) P: 2.43 Q: 2.49 R: 2.54 10. 8.5 11. (a) Any 2 of these: 1, 3, 5, 9, 15, 45 (b) 24, 48, 72... 12. (a) 10.41 (b) 15,336 13. 3560 + 2790 = **6350** The larger number is 6350.	14. 1242 in. ÷ 9 = **138 in.** The red ribbon is 138 in. long. 15. $1.25 x 6 = **$7.50** She spent $7.50. 16. 1.5 qt − 0.75 qt = **0.75 qt** 0.75 qt more water can be poured into the bottle. (Or 3 cups.) 17. 125 ml x 14 = 1750 ml = **1 ℓ 750 ml** He drinks 1 ℓ 750 ml in 2 weeks. 18. $11.90 + $27.35 = $39.25 $50 − $39.25 = **$10.75** He had $10.75 left. 19. $20.35 + $20.35 + $16.85 = **$57.55** They saved $57.55 altogether. 20. Paid by students: 18 x $3 = $54 Paid by Miss Chen: $72 − $54 = **$18** Miss Chen paid $18. 21. $2000 − $665 = $1335 $1335 ÷ 3 = **$445** Each microwave oven cost $445. 22. Rent for 2 months: $4500 x 2 = $9000 $9000 ÷ 4 = **$2250** Each person paid $2250.

Review	WB Review 4, pp. 90-93
Problem 5: Note that rounding the sum or product is not the same as finding the estimate. The answer is rounded. Problem 13: Rather than measuring the angles, they can be calculated.	

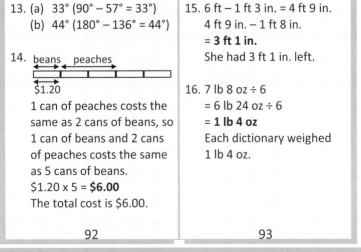

Review 4

1. (a) 10,590; 10,050; 9950; 9590; 9190
   (b) 8.3; 7.28; 2.83; 2.05

2. (a) 57.76
   (b) 4.43
   (c) 20.15
   (d) 282

3. (a) 51.2   (51.21)
   (b) 44   (43.65)

4. 26.08

5. A: 5.78 B: 5.84 C: 5.87

    90

6. 9 h 35 min

7. 1.7 m − 0.46 m = **1.24 m**

8. 10 x 48 = **480**

9. $\frac{6}{8} = \frac{3}{4}$

10. $\frac{8 \text{ hours}}{24 \text{ hours}} = \frac{1}{3}$

11. 4 gal − $\frac{3}{5}$ gal = **3$\frac{2}{5}$ gal**

12. $\frac{3}{8}$ of 40 = 3 x 5 = **15**

    91

13. (a) 33° (90° − 57° = 33°)
    (b) 44° (180° − 136° = 44°)

14. beans   peaches

    $1.20
    1 can of peaches costs the same as 2 cans of beans, so 1 can of beans and 2 cans of peaches costs the same as 5 cans of beans.
    $1.20 x 5 = **$6.00**
    The total cost is $6.00.

    92

15. 6 ft − 1 ft 3 in. = 4 ft 9 in.
    4 ft 9 in. − 1 ft 8 in.
    = **3 ft 1 in.**
    She had 3 ft 1 in. left.

16. 7 lb 8 oz ÷ 6
    = 6 lb 24 oz ÷ 6
    = **1 lb 4 oz**
    Each dictionary weighed 1 lb 4 oz.

    93

4   Symmetry

Objectives

- ♦ Identify figures with line symmetry.
- ♦ Identify lines of symmetry in geometric shapes.
- ♦ Complete a symmetric figure.

Suggested number of days: 5

		TB: Textbook WB: Workbook	Objectives	Material	Appendix
4.1	**Symmetric Figures**				
4.1a	Symmetry	TB: pp. 74-75 WB: pp. 99-100	♦ Identify figures with line symmetry.	♦ Small mirrors ♦ Paper ♦ Scissors	
4.1b	Lines of Symmetry	TB: pp. 76-79 WB: pp. 101-102	♦ Identify lines of symmetry in geometric shapes.	♦ Small mirrors ♦ Ruler ♦ Set-square	♦ Shapes (p. a22) ♦ Graph paper (p. a23) ♦ Square dot paper (p. a24)
4.1c	Symmetric Figures	TB: p. 79 WB: pp. 103-104	♦ Complete a symmetric figure.	♦ Ruler ♦ Set-square ♦ Small mirrors	♦ pp. a25-a26 ♦ Square dot paper (p. a24)
4.1d	Review	TB: pp. 83-85 WB: pp. 94-98	♦ Review.		

Blank Page

4.1 Symmetric Figures

Objectives

♦ Identify figures with line symmetry.
♦ Identify lines of symmetry in geometric shapes.
♦ Complete a symmetric figure.

Material

♦ Small mirrors with straight sides
♦ Sheets of paper and scissors
♦ Appendix pp. a22-a26
♦ Rulers
♦ Set-squares

Prerequisites

Students should know the names and basic properties of simple plane figures (squares, rectangles, triangles) and be able to identify parallel lines.

Notes

In earlier levels of *Primary Mathematics* students were introduced to plane figures and informally learned some of their properties by looking at right angles and parallel and perpendicular lines.

In this part students will further explore properties of plane figures by learning to identify figures with line symmetry. They will also learn the names and angle and side properties of some common quadrilaterals and triangles. The purpose here is to become familiar with these figures by looking at their symmetry, not to get bogged down with memorizing their names or properties. If you feel it is important that they learn and retain the names, you will have to spend extra time on vocabulary. In *Primary Mathematics* 5 they will spend more time with these figures and their properties and learn their names more thoroughly.

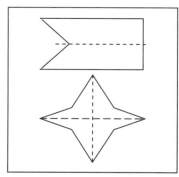

Plane figures that have a line of symmetry are called **symmetric figures**. The line of symmetry divides the figure into two parts, each of which is a reflection or mirror image of the other. When one part is flipped about the line of symmetry, it matches the other part. The two figures at the right are symmetric figures, and the dashed lines are **lines of symmetry**. Some figures, like the second one, have more than one line of symmetry. Students will not be required to find the total number of lines of symmetry in a figure at this level, but should recognize that there may be more than one line of symmetry.

Students will be investigating lines of symmetry in triangles and quadrilaterals.

An isosceles triangle is a triangle with two equal sides. It has one line of symmetry. An equilateral triangle is a triangle with three equal sides. It is a special case of an isosceles triangle and has three lines of symmetry. A triangle that has no sides equal is called a scalene triangle. A right-angled triangle is simply a triangle with a right angle. It can have two equal sides (isosceles) or none. In this curriculum students are not required to learn the word scalene until *Primary Mathematics* 5. You may teach it to them if you wish.

As you can see in the examples of the three types of triangles below, equal sides are indicated by marking them with a small cross hatch or notch.

Isosceles triangle
2 equal sides and 2 equal angles

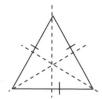

Equilateral triangle
3 equal sides and 3 equal angles

Scalene triangle
0 equal sides and 0 equal angles

A quadrilateral is a 4-sided figure. Students do not have to learn the term quadrilateral at this time for this curriculum, but you can teach it to them if you want.

A trapezoid is a 4-sided figure with only one pair of unequal sides. If a trapezoid is equilateral (the two non-parallel sides are equal) then it has one line of symmetry, otherwise it has no lines of symmetry.

A parallelogram is a 4-sided figure with both pairs of opposite sides parallel and equal in length.

A rhombus is a parallelogram with 4 equal sides.

A rectangle is a parallelogram with 4 right angles.

A square is a rectangle with 4 equal sides (so is also a rhombus).

For parallelograms, a rectangle and a rhombus have two lines of symmetry and a square has 4 lines of symmetry; all other parallelograms have no lines of symmetry.

In the figures below note that parallel lines are marked by small arrows; sides with the same number of arrows in a figure are parallel to each other. Again, notches are used to mark equal sides, and sides with the same number of notches in a figure are equal to each other.

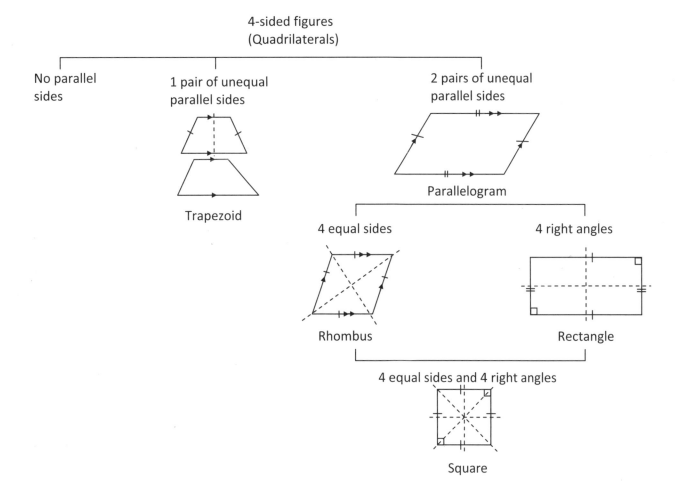

4.1a   Symmetry

Objectives

♦ Identify figures with line symmetry.

Vocabulary

♦ Symmetry
♦ Line of symmetry
♦ Reflection

Note

This is not likely to be a challenging lesson for students of any mathematical ability. It is a good opportunity to point out that art and drawing contains a lot of math concepts, and math allows for more precise and accurate drawings in creating patterns, such as in textiles, as well as in areas involving architecture or engineering.

There are many internet sites with projects and things to do with symmetry, if time permits interested students to explore them.

Identify symmetric figures	Text p. 74
Have students look at the figures on this page and ask them what they can tell you about the two halves of each figure. Each half is a mirror image or *reflection* of the other. If you have some small mirrors with a straight side, you can have students line up the mirror with the *line of symmetry* in each of the figures and see that the reflected image in the mirror is the same as the other half of the figure behind the mirror. There are many examples of symmetry in the environment, but often these examples are three dimensional, and the two halves are not exact mirror images. When various children's sites or books discuss line of symmetry for these objects, they are only really looking at the surface shape and pattern as if the surface were flattened, or copied onto the surface of a mirror. You may wish to tell students that solid objects are symmetric if you could cut them in half in such a way that one half is the "mirror image" of the other. Such shapes have "bilateral symmetry." With natural objects, such as animals, the two halves are not exactly identical, but they still are said to have bilateral symmetry if the two halves have the same parts. A person has bilateral symmetry.	
Activity	**Text p. 75, Tasks 1-2**
Get students to cut out some symmetric figures. The main concept they need to understand is that if you fold a figure on the line of symmetry, one side fits over the other exactly so that all the corresponding angles and sides are equal.	

Activity

Have students cut out complex figures from sheets of paper or index cards folded in half. Have them be careful to keep the pages they cut the figure out of as well as the figure intact. Collect both, but open the cut-out figure. Mix them up and change the orientation and put them up on the bulletin board or elsewhere. Ask students to match each cut-out with its source without moving them around.

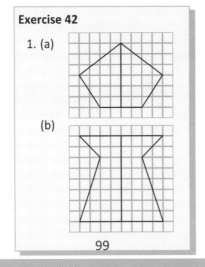

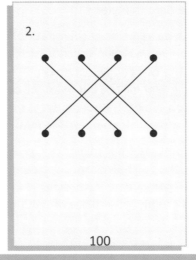

Objectives

♦ Identify lines of symmetry in geometric shapes.

Vocabulary

♦ Congruent
♦ Isosceles triangle
♦ Equilateral triangle
♦ Right-angled triangle
♦ Parallelogram
♦ Rhombus
♦ Trapezoid

Note

Students with good spatial sense may be able to determine if a line is a line of symmetry by visual inspection only. Most students will benefit from actually folding shapes, particularly the parallelograms or instances where the two halves are congruent, but not mirror images of each other.

It is up to you how much you want to emphasize the names of the shapes at this point. The vocabulary is covered in more detail in *Primary Mathematics* 5.

Discussion	Text pp. 76-78, Tasks 3-7
Provide students with paper cut-out rectangles, isosceles, equilateral, and scalene triangles, parallelograms, a rhombus, and two kinds of trapezoids to fold while you discuss these tasks. You can copy and cut out the figures on appendix p. a22.	3. Answers in textbook.

Task 3: Point out that for all three figures, the dotted line divides the rectangles into two halves that are exactly the same size and shape. When two shapes are the same, they are *congruent.* For 3(b), you can have students cut a rectangle along the diagonal to prove that the two halves are congruent. However, even if the two halves are congruent, the line dividing it is not necessarily a line of symmetry. It is only a line of symmetry if all parts match up when the figures are folded along it. A reflection of one half in a mirror set along the line of symmetry would look like the other half. If you have small mirrors, let students use them on the figures.

4. (a) Answer in textbook.
   (b) 3 possibilities

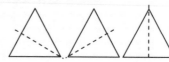

5. (c)

6. (a) no
   (b) yes

7. (a) yes
   (b) no

Task 4: Have students fold the triangle cut-outs to find lines of symmetry. Have them see how many lines of symmetry they can find for each of these triangles. There are 3 for the equilateral triangle, but only 1 for the isosceles triangle.

Task 5: Ask students why the triangle in (c) has a line of symmetry, but the ones in (a) and (b) do not. The one in (c) is an isosceles triangle, as well as a right triangle.

Task 6: Be sure students remember what parallel lines are. Parallel lines, if extended forever in either direction, will never cross each other. The lines on the two parallelograms in (a) and (b) do divide the shapes into congruent halves, but in (a) the dotted line is not a line of symmetry. You can have students see if they can find other lines of symmetry for parallelograms or rhombuses.

Task 7: Both the figures are trapezoids; they have one pair of parallel lines. In (a), the two halves are congruent; two of the sides are equal, but in (b) there are no equal sides. You can have students see if they can find other lines of symmetry for either kind of trapezoids.

Assessment	Text p. 79, Task 8
Also give students some paper cut-outs of squares. Ask them to find some lines of symmetry in a square. Then ask them in what ways a square is like a parallelogram, rectangle, or rhombus. A square is a parallelogram, rectangle, and rhombus.	8. (a) yes (b) yes (c) no (d) yes
Practice	WB Exercise 43, pp. 101-102

Activity

Have students draw some other symmetric plane figures using graph paper or square dot paper (appendix pp. a23-a24).

Enrichment

Students can find lines of symmetry by simple inspection, or by tracing the figure, cutting it out, and folding along the proposed line of symmetry. However, since they have learned in *Primary Mathematics* 4A how to draw perpendicular lines using a set-square and ruler, or on square grid paper, you can show them how to determine if a given line is a line of symmetry by having them draw perpendicular lines to the line of symmetry.

If a given line is a line of symmetry, then a line drawn from a point on one side to its corresponding point on the other side is perpendicular to the line of symmetry and is bisected (cut in half) by the line of symmetry. So if we can draw a line perpendicular to the proposed line of symmetry through a specific point on one side, such as a vertex, and this line does not go through the corresponding point on the other side, such as the opposite vertex, then the proposed line of symmetry is not a line of symmetry.

For example, in the parallelogram at the right, the sides on both sides of the dashed line look the same, but the dashed line is not a line of symmetry; a line perpendicular to the dashed line which goes through one corner does not go through the opposite corner.

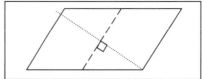

If students have transparent rulers, one can easily be placed perpendicular to the proposed lines of symmetry in the textbook by lining them up with one of the marks on the ruler, which are perpendicular to the sides of the ruler. If they do not have transparent rulers, students can use a set-square to put the ruler perpendicular to the proposed line of symmetry.

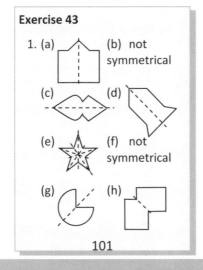

Exercise 43

1. (a)   (b) not symmetrical

(c)   (d)

(e)   (f) not symmetrical

(g)   (h)

101

2. (a) yes   (b) no

  (c) yes   (d) yes

  (e) no   (f) no

  (g) no   (h) yes

102

4.1c   Symmetric Figures

Objectives

♦ Complete a symmetric figure.

Note

If students have any trouble with determining if a figure is symmetric or completing a figure, have them cut the figure out and fold it along the proposed line of symmetry. Most students will be able to determine line symmetry by visual inspection, and be able to complete a figure just by comparing it to the other half visually. They can use small mirrors to check symmetry.

Discussion	Text p. 79, Task 8
Provide students with graph paper and have them copy the figure and then complete it using AB as the line of symmetry. They can then check their figures by folding along the line AB; they should be able to see their drawn lines through the paper. If not, they can cut out their figure and then see if it is symmetric.	8.
Assessment	**Appendix p. a25**
Have students complete the figures on appendix p. a25 using the dotted lines for lines of symmetry. For more challenge, you can have them complete the figures on appendix p. a26, or do that page after you discuss the enrichment activity on the next page of this guide.	

Activity

Students can draw figures using square graph paper or square dot paper for other students to complete. They need to be sure they have left room for the figures to be completed.

Enrichment

You can show students how to find where a corner or vertex of the figure should be by first drawing or imagining a perpendicular line from a vertex of the figure on one side to the line of symmetry, and then extending it the same distance on the other side. Some students may have discovered this on their own. Use the figure from Task 9 to show how this is possible, and then guide them in using this approach with the first figure on appendix p. a26, as shown at the right. Then let them complete the rest of the figures on that page.

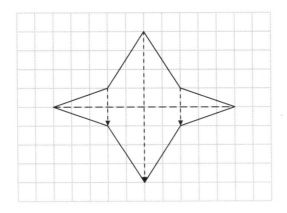

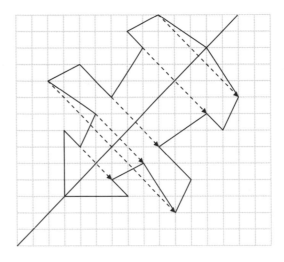

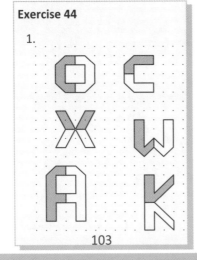

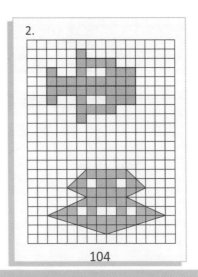

Objectives

♦ Review.

Note

In the workbook there are two reviews after the last exercise for Unit 3, but in the textbook there are two reviews after Unit 4. Answers were provided for the first of the two textbook and workbook reviews on pp. 126 -127 of this guide.

Review D	**Text pp. 83-85**

Review D

1. (a) 123.58, 132.85, 135.28, 251.83
   (b) 123.58: 20
   132.85: 2
   135.28: 0.2
   251.83: 200
   (c) 123.58

2. (a) 6
   (b) 7

3. (a) 450.07   (b) 35.53
   (c) 30.54   (d) 107.08

4. 42

5. (a) 27,360   (b) 8262   (c) 32,103

6. (a) $1\frac{4}{9}$ (b) $1\frac{3}{8}$ (c) $1\frac{2}{5}$

   (d) $\frac{1}{8}$ (e) 50   (f) $4\frac{1}{2}$

7. (a) $1\frac{9}{25}$

   (b) 3.22

8. $12.25

9. (a) Perimeter = 26 cm   (b) Perimeter = 62 cm
   Area = 22 cm^2 Area = 138 cm^2

10. (a)

	Walkathon	Swimming Competition	Carnival
Boys	56	63	45
Girls	50	66	47

   (b) 327

11. length + width = 42 in. ÷ 2 = 21 in.
    width = 21 in. − 12 in. = **9 in.**
    The rectangle has a width of 9 in.

Text pp. 83-85

12. side = 5 m   (5 x 5 = 25)
    perimeter = 5 m x 4 = **20 m**
    The flower bed has a perimeter of 20 m.

13. (a) 340 ml
    (b) 744 g
    (c) 5 in.

14. (a) 7 h 35 min
    (b) 10:05 p.m.

15. Cost of picture book:
    $\frac{2}{5}$ of $34 = 2 x $\frac{\overset{\$34}{34}}{5}$ = 2 x $6.80 = **$13.60**

    Total spent: $8.25 + $13.60 = **$21.85**
    He spent $21.85 altogether.

16.

    2 units = $14
    1 unit = $14 ÷ 2 = $7
    3 units = $7 x 3 = **$21**

    Or: $\frac{2}{5}$ of total = $14

    $\frac{1}{5}$ of total = $14 ÷ 2 = $7

    $\frac{3}{5}$ of total = $7 x 3 = $21

    The tennis racket cost $21.

Review	WB Review 5, pp. 94-98
Q 3: This question can be confusing since students have not learned order of operations and so do not know that multiplication is done before addition. You many want to rewrite the problem as 7 x 8 = _____ x 8 + 8 + 8. On the right hand side, 5 eights + 2 more eights equals the 7 eights of the left-hand side.	

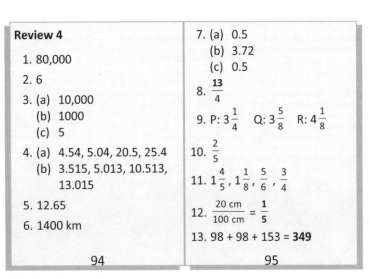

Review 4

1. 80,000

2. 6

3. (a) 10,000
   (b) 1000
   (c) 5

4. (a) 4.54, 5.04, 20.5, 25.4
   (b) 3.515, 5.013, 10.513, 13.015

5. 12.65

6. 1400 km

94

7. (a) 0.5
   (b) 3.72
   (c) 0.5

8. $\frac{13}{4}$

9. P: $3\frac{1}{4}$ Q: $3\frac{5}{8}$ R: $4\frac{1}{8}$

10. $\frac{2}{5}$

11. $1\frac{4}{5}, 1\frac{1}{8}, \frac{5}{6}, \frac{3}{4}$

12. $\frac{20\ cm}{100\ cm} = \frac{1}{5}$

13. 98 + 98 + 153 = **349**

95

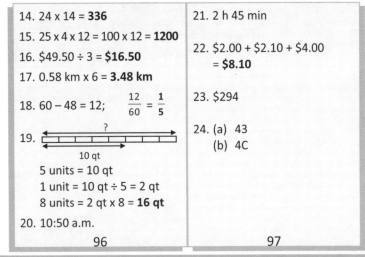

14. 24 x 14 = **336**

15. 25 x 4 x 12 = 100 x 12 = **1200**

16. $49.50 ÷ 3 = **$16.50**

17. 0.58 km x 6 = **3.48 km**

18. 60 − 48 = 12; $\frac{12}{60} = \frac{1}{5}$

19.

   10 qt
   5 units = 10 qt
   1 unit = 10 qt ÷ 5 = 2 qt
   8 units = 2 qt x 8 = **16 qt**

20. 10:50 a.m.

96

21. 2 h 45 min

22. $2.00 + $2.10 + $4.00
    = **$8.10**

23. $294

24. (a) 43
    (b) 4C

97

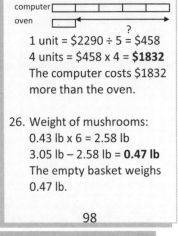

25.

   $2290
   computer
   oven
   ?
   1 unit = $2290 ÷ 5 = $458
   4 units = $458 x 4 = **$1832**
   The computer costs $1832 more than the oven.

26. Weight of mushrooms:
    0.43 lb x 6 = 2.58 lb
    3.05 lb − 2.58 lb = **0.47 lb**
    The empty basket weighs 0.47 lb.

98

5 Solid Figures

Objectives

- Interpret 2-dimensional drawings of unit cubes.
- Build solids from models drawn on isometric dot paper.
- Determine the number of unit cubes used to make a solid.
- Visualize new solids formed by adding or removing cubes.

Suggested number of days: 3

		TB: Textbook WB: Workbook	Objectives	Material	Appendix
5.1	**Identifying Solid Figures**				
5.1a	Unit Cubes	TB: pp. 86-87 WB: pp. 105-106	◆ Interpret 2-dimensional drawings of unit cubes. ◆ Determine the number of unit cubes used to make a solid.	◆ Centimeter cubes ◆ Multilink cubes	◆ Isometric dot paper (p. a27) ◆ p. a28
5.1b	Solids from Models	TB: pp. 88-89 WB: pp. 107-108	◆ Build solids from models drawn on isometric dot paper.	◆ Multilink cubes ◆ Box (see lesson)	◆ Isometric dot paper (p. a27) ◆ p. a29
5.1c	Remove or Add Cubes	TB: p. 89 WB: pp. 109-110	◆ Visualize new solids formed by adding or removing cubes.	◆ Multilink cubes	◆ p. a30

Blank Page

5.1 Identifying Solid Figures

Objectives

- Interpret 2-dimensional drawings of unit cubes.
- Determine the number of unit cubes used to make a solid.
- Build solids from models drawn on isometric dot paper.
- Visualize new solids formed by adding or removing cubes.

Material

- Centimeter cubes or other cubes
- Multilink cubes
- A box shape made from multilink cubes with surfaces covered with paper.
- Appendix pp. a27-a30

Prerequisites

Students should be familiar with the shape of cubes and other rectangular prisms.

Notes

In this part students will learn to visualize two-dimensional representations of simple solids made out of unit cubes. This is a precursor to the next part, where they will learn that volume is the amount of space a solid occupies and is measured in cubic units. In this part they will find the number of cubes used to create a solid drawn in two dimensions, such as those shown below. They need to be able to interpret a solid made up of unit cubes drawn in two dimensions as a 3-dimensional object in order to later understand similar drawings used to explain the volume of rectangular prisms (boxes) and cubes. The drawings will include cubes they can't see but will have to count. They will first explore solids drawn in 2 dimensions with the aid of isometric dot paper and build solids to match the drawings.

Students can use unit cubes of any size, such as centimeter cubes, that do not link to each other, or multilink cubes that do. However, in all of these exercises, assume that the figures drawn on the isometric dot paper can be constructed from blocks that do not link. That is, when a block is not on the lowest level, one or more blocks have to be under it so that no block is suspended in the air. For example, in the pyramid shape to the right, there are blocks that are hidden behind other blocks. This figure needs to have 10 blocks in order to construct it, 4 of which are hidden in the drawing, but which need to be there to support other blocks.

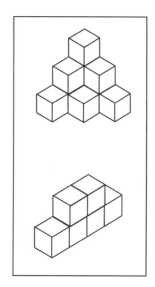

Students must also assume that there are hidden blocks **only** when one is necessary to support another block. For example, in the 6-block figure at the right, while there could be a block hiding behind the second to the last block in the bottom row, such a block would not be a supporting block, so we know the figure has just the 6 blocks that we see.

If you do have students use multilink cubes, be sure they are capable of ignoring the protrusions and holes and "seeing" just the cubical shape when comparing them to drawings. Be sure they understand that the figures in the drawings can be built of cubes that do not stay stuck together.

If students use cubes that do not link, make sure they understand that for all the solids in this part of the textbook and workbook the cubes that are next to each other touch over their entire opposing faces (as if they did link together).

In this curriculum, the term "cuboid" refers to a rectangular prism, or box shape.

Blank Page

Objectives

♦ Interpret 2-dimensional drawings of unit cubes.

♦ Determine the number of unit cubes used to make a solid.

Note

Building 3-dimensional models from 2-dimensional pictures can be challenging for some students. Allow them sufficient time to build cubes based on the drawings. Other students may find this lesson quite easy. You can have them spend more time trying to draw cubes or represent their own builds on isometric dot paper.

Draw cubes	
Give each student a cube and have them look at it from different angles. Tell them that the cube takes up space. Point to three of the sides of the cube that go in different directions and tell them that the three directions define the shape of the cube. Tell them that if we want to draw a cube on paper, we try to make it look like a real cube, even though the paper is flat. Ask them to attempt drawing the cube on paper. You may want to demonstrate some ways to draw a cube on paper. To draw a cube with a face front-most, start by drawing a square, one face of the cube. Draw parallel lines of equal length backwards and to the side from three of the corners. Connect the ends of these lines with lines parallel to the face. To draw a cube with an edge front-most, draw two dots a little ways apart horizontally, and two dots between and just above and below the first two. Connect the dots into a diamond shape. This is the top face of the cube. Draw parallel lines of the same length straight down from the two ends and from the near corner. Connect the ends of the lines.	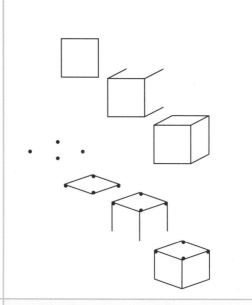
Discussion	**Text pp. 86-87, Tasks 1-3**
Tell students that the cubes drawn on both sides of page 86 are drawn to look like real cubes. The ones on the left, the alphabet blocks, look a bit more real because of shading. The ones on the right are also drawings of cubes, the type they will typically see in a math book. Task 1: Provide students with more cubes. Tell them that the dots on the paper help to draw lots of cubes all the same size. Have them put two cubes together and compare them to the pictures. Task 2(a): Make sure students realize that the solid is made up of 4 cubes, even though only three are showing. There is a hidden cube in the back. Without it, the figure is not possible.	

Task 2(b): Tell students that for the problems they do in the textbook and workbook, the faces of the solids they build must touch along the whole face.

Have students compare their solids. Point out that however they are built, they all take up the same amount of space equivalent to 4 cubes.

Task 3: Be sure that students understand that they need to build a cube out of the 8 smaller cubes. A drawing of one is shown at the right.

After they have finished Task 3, have them build other solids of different shapes with the 8 cubes and compare. Again, each different shape takes up the same amount of space.

Assessment	Appendix p. a28
Give students copies of appendix p. a28. Have them first try to determine how many cubes are used for each figure. Then have them build the solids.	Number of cubes for each solid: A: 6 C: 11 B: 6 D: 10 E: 15 F: 14 G: 10
Practice	WB Exercise 45, pp. 105-106
Q 2: The solids for this page do not show lines separating individual cubes. If you think this will be an issue with students, do this question or exercise after the next lesson.	

Activity/Enrichment

Have students draw some of the figures they build with 8 blocks using isometric dot paper (appendix p. a27). For more challenge, let them draw solids they build from more blocks. Other students could try to build the solids they draw.

Exercise 45

1. Check students' builds.

   (a) 3

   (b) 5

   (c) 7

105

2. Check students' builds.

106

5.1b   Solids from Models

Objectives

♦ Build solids from models drawn on isometric dot paper.

Vocabulary

♦ Cuboid
♦ Rectangular prism

Note

In this lesson the cubes are not drawn individually; that is, only the lines for the outside edges of the figures remain. Allow students plenty of practice building the solids depicted in the pictures. More advanced students can attempt drawing some solids they build in the same way, where they only draw the outside edges of the figures.

Solid figures	
Before the lesson build a rectangular prism, or box shape, from cubes. Cut paper the same size as each surface and tape the paper to the sides so that students do not see the individual blocks. Show students your creation and an individual block. Have them estimate the number of blocks you used to build the shape. Then remove the paper sides so they can see the individual cubes and have them count the cubes.	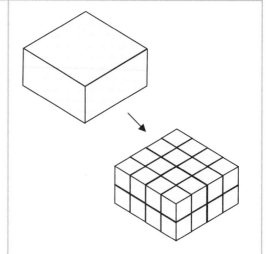
Discussion	**Text pp. 88-89, Tasks 4-6**

Task 4: Tell students that a box shape is sometimes called a *cuboid*. Later they will learn that it is more commonly called a *rectangular prism*. The one pictured in this task can be built out of cubes. Have students select the correct number of cubes (16) and build the solid from the unit cubes.
 	
Tasks 5-6: Note that solids D, P, and Q each have a "hidden" cube.
 	
Have students build the solids and write down their answers for how many cubes are needed for each. If there is a time constraint students can work in groups and each build one or two of the solids, but if students need more experience let them all build each one.
 	
Ask students which of the 6 solids in these two tasks take up the same amount of space. (C, D, and P) Which one takes up the least amount of space? (A)
	4. 16
	
	5. A: 5
	  B: 6
	  C: 9
	  D: 9
	
	6: P: 9
	  Q: 8

Assessment	Appendix p. a29
Give students copies of appendix p. a29. Have them first try to determine how many cubes are used for each figure. Then have them build the solids.	Number of cubes for each solid: A: 24 B: 14 C: 7 D: 13 E: 20 F: 52
Practice	WB Exercise 46, pp. 107-108

Activity/Enrichment

Have students draw some solids they build using isometric dot paper, only drawing the outside lines. Other students could try to build the solids they draw.

Exercise 46

1.

Solid	Number of unit cubes
A	3
B	4
C	2
D	6
E	5
F	8

107

2.

Solid	A	B	C	D	E
Number of unit cubes	16	27	6	9	7

108

5.1c Remove or Add Cubes

Objectives

♦ Visualize new solids formed by adding or removing cubes.

Note

In this lesson students will look at two drawings and compare to see how many cubes have been added or removed. Some students will have to build the solids. Other students may be able to determine the change by simply looking at the drawings. You can have these students use isometric dot paper to draw similar problems for other students to solve.

Change in solid figures	
Before the lesson use multilink cubes to build two or more solids, similar to each other but with some cubes added or removed. You can make several versions. Show the solids to students and ask them for the difference in number of cubes between them. Verify their answers by letting a student add or remove cubes as needed. When comparing two solids, ask them which one takes up more space and by how many cubes.	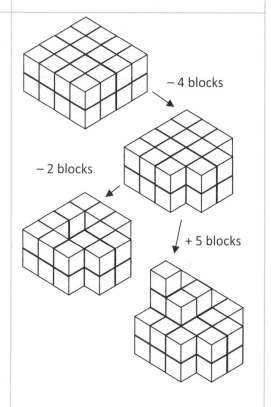
Discussion	**Text p. 89, Tasks 7-8**

Task 7: Have students build the first solid, and then remove cubes to form the second solid. Ask them which of the two solids takes up the most space.
 	
Task 8: Have students build the first solid, and then add cubes to form the second solid. Again ask which of the two solids takes up the most space.
	7. A: 12 B: 10
	    2 cubes are removed.
	
	8. C: 4 D: 6
	    2 cubes are added.

Assessment	Appendix p. a30
Give students copies of appendix p. a30. Have them build the solids on the left side of the page and either add or remove cubes to make the solids on the right side. They should write down the total number of cubes needed for each solid and how many were added or removed. You may want to ask them to first find the answers before building the solids.	A: 24 $\xrightarrow{-6}$ 18 B: 11 $\xrightarrow{+3}$ 14 C: 11 $\xrightarrow{+10}$ 21
Practice	WB Exercise 47, pp. 109-110

Enrichment - Surface Area

You can have students explore surface area of the solids they can build from cubes. Remind them that area is measured in square units. Hold up a cube and ask students for the surface area of the cube, that is, the area on the surface of the cube. It is 6 square units, one for each space. Then do the same with a rectangular prism made from multilink cubes and with some other shapes. You can tell them to imagine the solid is dipped in paint. How many faces would the paint cover? If you have the resources, you could do just that — dip a solid into paint, let the paint dry, and have students determine how many faces are covered with paint.

Have students build shapes and find the surface areas. Have them see if they can find the surface area of any of the shapes in the textbook tasks or workbook exercises, and then build the shapes and count to verify.

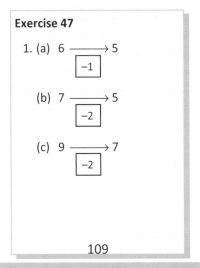

Exercise 47

1. (a) 6 $\xrightarrow{\boxed{-1}}$ 5

   (b) 7 $\xrightarrow{\boxed{-2}}$ 5

   (c) 9 $\xrightarrow{\boxed{-2}}$ 7

109

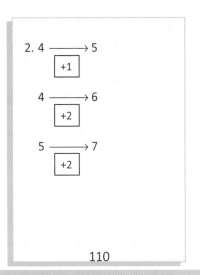

2. 4 $\xrightarrow{\boxed{+1}}$ 5

   4 $\xrightarrow{\boxed{+2}}$ 6

   5 $\xrightarrow{\boxed{+2}}$ 7

110

6 Volume

Objectives

♦ Find the volume of a solid in cubic units and cubic centimeters.
♦ Learn other units of volume and compare relative sizes.
♦ Use a formula to find the volume of a cuboid (rectangular prism).
♦ Convert between cubic centimeters and liters.

Suggested number of days: 8

		TB: Textbook WB: Workbook	Objectives	Material	Appendix
6.1	**Cubic Units**				
6.1a	Cubic Units	TB: pp. 90-91 WB: p. 111	♦ Find the volume of a solid in cubic units. ♦ Understand the cubic centimeter as a unit of volume.	♦ Cubes ♦ Centimeter cubes ♦ Boxes	
6.1b	Cubic Centimeters	TB: pp. 91-92 WB: p. 112	♦ Find the volume of a solid in cubic centimeters. ♦ Learn other units of volume and compare relative sizes.	♦ Multilink cubes ♦ Centimeter cubes ♦ Inch cube ♦ Foot cube ♦ Meter sticks	♦ pp. a29-a31
6.2	**Volume of a Cuboid**				
6.2a	Volume of a Cuboid	TB: pp. 93-96 WB: pp. 113-114	♦ Use a formula to find the volume of a cuboid (rectangular prism).	♦ Cubes	
6.2b	Practice	TB: p. 98	♦ Practice.	♦ Inch cube ♦ Foot cube	♦ p. a32
6.2c	Cubic Centimeters and Liters	TB: pp 97, 99 WB: pp. 115-116	♦ Convert between cubic centimeters and liters.	♦ Measuring cup ♦ 1000-cube ♦ Dropper or teaspoon	
6.2d	Review	TB: pp. 100-104 WB: pp. 117-128	♦ Review.		

Cubic Units and Volume of a Cuboid

Objectives

♦ Find the volume of a solid in cubic units and cubic centimeters.
♦ Learn other units of volume and compare relative sizes.
♦ Use a formula to find the volume of a cuboid (rectangular prism).
♦ Convert between cubic centimeters and liters.

Material

♦ Cubes such as multilink cubes
♦ Centimeter cubes
♦ Boxes of various shapes to fill with cubes
♦ Inch cube (Can use net on appendix p. a31)
♦ Foot cube (Make from 6 squares of cardboard)
♦ Meter sticks
♦ Appendix pp. a29-a32
♦ Liter measuring cup
♦ 1000-cube from base-10 set
♦ Dropper or teaspoon

Prerequisites

Students should be able to interpret 2-dimensional drawings of cubes and count the number of cubes in those drawings, even if some are hidden.

Notes

In the previous unit students counted the number of cubes used to make a solid figure. They were essentially finding the volume in cubic units. In this unit they will be introduced to the standard units for volume, primarily the cubic centimeter, and their abbreviations (e.g., cm^3, m^3, $in.^3$, ft^3, yd^3).

In *Primary Mathematics* 3B students learned to find the area of a rectangle given its length and width. This was reviewed in *Primary Mathematics* 4A and students found the area of compound figures. In this unit students will learn to find the volume of a cuboid (rectangular prism) by multiplying its length by its width by its height.

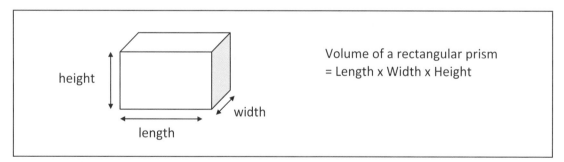

Volume of a rectangular prism
= Length x Width x Height

In *Primary Mathematics* 2 and 3 students learned to find the capacity of containers in liters and milliliters. In this unit students will learn that liters and milliliters are also used to measure the volume of liquid in a container, not just the capacity of the whole container. Since they have used measuring cups or beakers to determine the capacity of other containers, this is not really a new concept. However, they will learn that 1 cm^3 is equivalent to 1 milliliter, and 1000 cm^3 is equivalent to 1 liter, and to convert cubic centimeters to liters and milliliters and vice versa.

Volume is also measured in cups, quarts, and gallons, such as using a measuring cup to measure out a volume of liquid to add to a recipe, or how many gallons of gasoline to add to the car tank. As students advance in science, the metric system will be used more often than the U.S. customary system.

Although the liter is the more common unit for volume, the international standard unit of volume is the cubic meter, which is equal to 1000 liters.

Blank Page

6.1a   Cubic Units

Objectives

♦ Find the volume of a solid in cubic units.
♦ Understand the cubic centimeter as a unit of volume.

Vocabulary

♦ Dimensions
♦ Volume
♦ Cubic unit
♦ Cubic centimeter

Note

Going from cubic units to cubic centimeters as a standard unit of measurement will not likely be a difficult concept.

If you choose not to have students fill boxes with cubes, you can combine this lesson with the next.

Introduce cubic units and centimeters

Point to the straight edge of some item and ask how we measure its length. We use standard units, such as centimeters. Draw a line on the board and label the line as 1 cm. Now point to a flat surface and ask how we measure its area. We measure it by how many squares would fit over it. Draw a square and label two sides 1 cm. Remind students that we call this a square centimeter and write it as cm^2. Two directions, or *dimensions*, define its size. Remind them that the object we are measuring does not also have to fit squares exactly; we can use parts of squares along curved surfaces and the area would be the same as those parts put into whole squares.

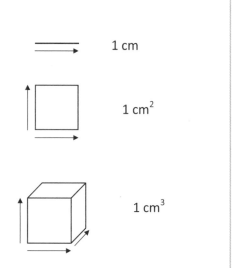

Point to a solid object and say that the object takes up space. The amount of space it takes up is called its *volume*. Ask them how we might measure volume. We measure the space something takes up in *cubic units*. Show them a centimeter cube. Tell them that as with length and area, we use standard measurements, and one standard measurement for volume is a cube 1 centimeter on each side. We call this a *cubic centimeter,* and write it as cm^3. Three directions, or dimensions, define its size. Draw a cube on the board and label 3 sides as 1 cm. Remind them that the picture is just a way of showing a solid on a flat piece of paper and we try to make it look 3-dimensional.

Show students a small box and ask some of them to fill it in with cubes to determine its approximate volume. You can use cubes of any size as long as they are the same size, since this lesson is on cubic units rather than standard units. If you have enough, give each group a box. They do not all have to have straight sides. The point that will be made is that we measure the volume in cubic units even if it is not a rectilinear figure.

Ask students to imagine how we could measure the volume when cubes don't fit in the space exactly. We could cut up some cubes into bits and pieces to fill in the extra spaces. When we say an object has a volume of 1000 cubic units, for example, then the amount of space it takes up is the same as the amount of space 1000 unit-cubes take up even if the shape of the space is not the same. One way to illustrate this is to use some clay shaped as a cube. Say the volume is 1000 cubic units and then deform it. The volume is still 1000 cubic units.	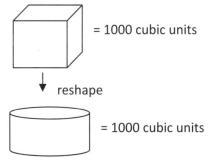
Discussion	**Text p. 90, Task 1**
Page 90: If necessary have students build the solids shown on this page to verify they all use the same number of cubes. They all have the same volume.	1. 6 cubic units
Assessment	**Text p. 91, Task 2**
Have students find the volumes without building the solids. After they complete this task discuss the strategies they might have used to find the number of cubes for C and D. They could have simply imagined the lower layers and counted the hidden cubes, or multiplied the number of cubes in one layer by the number of layers.	2. A: 5 cubic units B: 9 cubic units C: 18 cubic units D: 12 cubic units
Practice	WB Exercise 48, pp. 111

Exercise 48

1. A: 12   B: 6

   C: 16   D: 15

   E: 10   F: 15

   C

   B

   111

6.1b   Cubic Centimeters

Objectives

♦ Find the volume of a solid in cubic centimeters.
♦ Learn other units of volume and compare relative sizes.

Vocabulary

♦ Cubic meter
♦ Cubic inch
♦ Cubic foot
♦ Cubic yard

Note

Although students will usually be measuring in cubic centimeters, it is important for them to know other standard units for measuring volume, to have a feel for their relative sizes, and to see how much volume increases when the length increases by the same amount on each side.

Discussion	Text pp. 91-92, Tasks 3-5
Task 3: This task emphasizes the abbreviation used to show cubic units. Provide students with centimeter cubes and have them actually do this task so they get a feel for the size of a cubic centimeter.	3. The volume of the solid at the top of p. 92 is **3** cm^3. **6** cubes were used to build the second solid. The volume of the third solid is **6** cm^3.
Task 4: Ask students how they found the volume and discuss ways to find the volume of this solid without actually counting each and every cube. We could count the cubes along each side, multiply them together to get the number of cubes in the top layer (4 x 4 = 16) and double that to get the number of cubes in both layers (16 x 2 = 32).	4. 32 cm^3 5. A: 4 cm^3 B: 12 cm^3 C: 10 cm^3 D: 12 cm^3
Task 5: You can also ask students which solid has the largest volume and what the difference in volume is between several of the solids. Ask them how they can find the difference in volume without finding the actual volumes of the two solids. For example, to get from solid C to solid D we could move one cube down to the bottom layer and add 2 cubes on the end, so 2 cubes were added and the difference in volume is 2 cm^3.	
Assessment	Appendix pp. a29 or a30
Tell students that the cubes used are all centimeter cubes and ask them to write down the volume of each solid without actually building them.	
Volume of solids made from larger cubes	
Show students a multilink cube (2 cm on a side) and a centimeter cube. Put them next to each other so they can compare the size and tell them that the multilink cube is 2 cm on each side. Ask them what its volume is in cubic centimeters. You can have a student build a cube the same size as the multilink cube using the 1-cm cubes. 8 cubes are needed and the volume is 8 cm^3.	
Have students build larger cubes or other shapes with the multilink cubes and find the volume in cubic centimeters. For	

example, if they build a cube with 8 multilink cubes, the volume is 64 cm^3. Each multilink cube has a volume of 8 cm^3, so 8 of them have a volume of 64 cm^3.

Other units for volume

Show students a cube that is an inch along each side. You can use the net on appendix p. a31 to make a paper cube if you do not have any inch cubes. Have a student measure the sides and tell them that the volume is 1 cubic inch, or 1 in.^{3}. A volume of 1 in.^{3} is as big as a cube with side 1 in. Have them compare its size to a cubic centimeter block. Something that has a volume of, say, 10 in.^{3} will be quite a bit larger than something that has a volume of 10 cm^3.

Tape some paper together to make a similar net with 6 square feet and make a cube from it. The volume is 1 cubic foot, or 1 ft^3. Have students compare its volume visually with a cubic inch. Save these two paper cubes for an activity in the next unit.

Either make a box that is 1 meter square or do the following: Use masking tape to mark a square with 1-meter sides on the floor with one side against the wall. Mark another square meter on the wall with the same edge as the square on the floor. Use meter sticks to mark two opposite vertical edges. Tell students to imagine a cube of this size. Its volume is 1 cubic meter, or 1 m^3. Have them compare its volume visually with a cubic centimeter.

You can do something similar for a cubic yard, using yard sticks, or point out how a cubic yard compares to a cubic meter; they are similar in size.

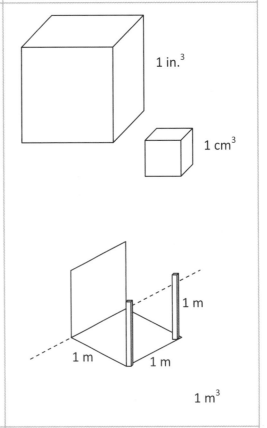

Practice	WB Exercise 49, pp. 112

Exercise 49

1. (a) 6 cm^3 (b) 6 cm^3

    (c) 18 cm^3 (d) 16 cm^3

    (e) 6 cm^3 (f) 9 cm^3

112

6.2a Volume of a Cuboid

Objectives

♦ Use a formula to find the volume of a cuboid (rectangular prism).

Note

This lesson should seem a natural extension of the previous lessons. Students may have already derived the formula for a cuboid or rectangular prism from the experience with cubes. Finding the volume given just the measurements for each side is more abstract than seeing the cubes.

Introduce formula	
Use multilink cubes or other cubes. Form a rectangle with a single layer of 4 cubes by 2 cubes. Ask students for its volume. Point out that we can find the volume by multiplying the number of cubes along the length by the number of cubes along the width. Also point out that the height is 1 unit.	 Number of cubes: 4 x 2 x 1 = 8
Add another layer and ask for the volume. Since we know how much is in one layer, we can find the number in both layers by multiplying the number of cubes on one layer by 2.	Number of cubes: 4 x 2 x 2 = 16
Add another layer and ask for the volume. There are now 3 layers with 4 x 2 in each layer.	Number of cubes: 4 x 2 x 3 = 24
The length is 4 units, the width is 2 units, and the height is 3 units. We can find the volume by multiplying these measurements together.	Volume = length x width x height = 4 units x 2 units x 3 units = 24 cubic units
Point out that the order in which we multiply the sides, i.e., which side we take as the first layer, does not matter. It could be the layer on one side or the front instead of the bottom or top.	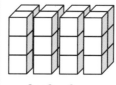 2 x 3 = 6 6 x 4 = 24 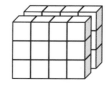 4 x 3 = 12 12 x 2 = 24
For example, if we had a height of 5 cubes, we could multiply the height by the width first. 5 x 2 x 4 = 10 x 4 = 40. This is a little easier to calculate mentally than 4 x 2 x 5 = 8 x 5 = 40. You can repeat with other examples.	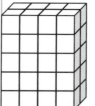 5 x 2 = 10 10 x 4 = 40 Volume = 40 cubic units

Discussion	Text pp. 93-95, Tasks 1-3

Page 93: This page shows pictorially what you already showed students concretely. The bottom of the page shows that we don't have to see the individual unit cubes to be able to determine the volume. If we know the measurements for length, width, and height, we can just multiply them together to get the volume.
 	
Tasks 2-3: Point out that these pictures are not actual size; they are drawn to scale and the actual sizes of these figures will be larger than shown here and more different from each other as appears on this page.
	24 cm^3
	
	1. (a) 3 cm (b) 5 cm
	    2 cm 3 cm
	    2 cm 1 cm
	    12 cm^3 15 cm^3
	
	(c) 3 cm (d) 4 cm
	    3 cm 2 cm
	    3 cm 5 cm
	    27 cm^3 40 cm^3
	
	2. 80; 80 in.^3
	
	3. 2; 2 m^3

Assessment	**Text p. 96, Tasks 4-6**
	4. 60 m^3 5. 27 m^3 6. A: 54 cm^3 B: $30,000 \text{ cm}^3$ C: 350 m^3 D: 180 m^3
Practice	WB Exercise 50, pp. 113-114

Exercise 46

1.

Solid	Length	Width	Height	Volume
B	2 in.	2 in.	2 in.	8 in.^3
C	5 in.	2 in.	4 in.	40 in.^3
D	3 in.	2 in.	7 in.	42 in.^3
E	7 in.	3 in.	2 in.	42 in.^3

113

2. 18 cm^3

200 cm^3

126 cm^3

192 m^3

240 m^3

114

Objectives

♦ Practice.

Note

Practice 6A on p. 98 can be done before or after text p. 97. It is given here before the next lesson to allow students more time to consolidate the concepts just learned.

Practice	Text p. 98, Practice 6A
	1. (a) 9 in.^{3} (b) 15 in.^{3}
	2. 30 cm x 25 cm x 15 cm = **11,250 cm^3**
	3. 5 cm x 5 cm x 5 cm = **125 cm^3**
	4. 12 ft x 10 ft x 3 ft = **360 ft^3**
	5. 8 x 5 x 3 = **120 cubes**
	6. 30 cm x 20 cm x 20 cm = **12,000 cm^3**
Assessment	**Appendix p. a32**

Give students copies of appendix p. a32 or write the problems on the board. After students work on them, have them discuss their solutions.

1. Find the volume of a box with a length of 15 cm, a width of 12 cm, and a height of 4 cm.

2. Find the volume of a box that is 11 cm by 2 cm by 8 cm.

3. Find the volume of a 5-centimeter cube.

4. A rectangular prism is made from 2-centimeter cubes. Its dimensions are 10 cubes by 8 cubes by 4 cubes. What is its volume?

   We can find the volume of the total number of cubes, then the volume of 1 cube, and multiply. Or we can first find the dimensions of each side and then the volume.

5. A rectangular container is 11 cm long, 11 cm wide, and 9 cm high. How many 2-centimeter cubes can it hold?

   We cannot simply find the total volume and divide by the volume of a 2-cm cube. If we put a row in the bottom layer, we will get a gap of 1 cm along both the length and width where we cannot fit a whole cube. So we need to first find how many cubes can fit along the length, width, and height.

1. 15 cm x 4 cm x 12 cm = 720 cm^3

2. 2 cm x 8 cm x 11 cm = 176 cm^3

3. 5 cm x 5 cm x 5 cm = 125 cm^3

4. Total cubes: 10 x 8 x 4 = 320
   Each cube is 8 cm^3.
   8 cm^3 x 320 = 2560 cm^3
   Or: 10 x 2 cm = 20 cm
   8 x 2 cm = 16 cm
   4 x 2 cm = 8 cm
   20 cm x 16 cm x 8 cm = 2560 cm^3

5. 5 cubes will fit along the length and width, and 4 along the height.
   Total number of cubes: 5 x 5 x 4 = 100

Enrichment

Draw a cube and label the edges of the cube as 1 ft. Ask students to find the volume in cubic inches. Point out that just because there are 12 inches in a foot, we cannot say there are 12 cubic inches in a cubic foot. If you made the paper foot and inch cubes in the earlier enrichment activity, put the cubic inch inside the cubic foot. They should see that a lot more than 12 will fit. To find the volume of a cubic foot in cubic inches, we need to change the measurement units for each side to inches, and then use that to find the volume in cubic inches. There are 1728 cubic inches in a cubic foot.

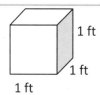

Volume of 1 ft^3 in $in.^3$:
1 ft x 1 ft x 1 ft = 1 ft^3
1 ft = 12 in.
12 in. x 12 in. x 12 in. = 1728 $in.^3$
1 ft^3 = 1728 $in.^3$

Draw another cube or re-label the edges of the previous drawing with 1 yd. Ask students to find the volume in cubic feet. There are 27 cubic feet in a cubic yard.

Volume of 1 yd^3 in ft^3:
1 yd x 1 yd x 1 yd = 1 yd^3
1 yd = 3 ft
3 ft x 3 ft x 3 ft = 27 ft^3
1 yd^3 = 27 ft^3

Similarly, draw a cube, label the edges 1 m, and guide students in calculating the number of centimeters in a cubic meter. There are 1 million cubic centimeters in a cubic meter.

Volume of 1 m^3 in cm^3:
100 cm x 100 cm x 100 cm
= 1,000,000 cm^3

6.2c   Cubic Centimeters and Liters

Objectives

♦ Convert between cubic centimeters and liters.

Note

Since the cube in a base-10 set is approximately a centimeter cube, the 1000-cube is approximately the same volume as 1000 cm^3, or 1 liter. Some supply stores do carry liter cubes that can hold water; if you have one you can use that instead.

The idea that 1000 cm^3 is not the same as 1 m^3 is an important one to stress.

Relate cubic centimeters to milliliters	
Show students a measuring cup with 1 liter of water. If you have a dropper, use it to squeeze out about 20 drops into a small container. Tell them that this is 1 milliliter. If you do not have a dropper, measure out a teaspoon of water into another container and tell them that this is about 5 milliliters. Ask them how many milliliters are in a liter. There are 1000 milliliters in a liter.	1 ℓ = 1000 ml 1 ml = 1 cm^3
Show students a centimeter cube. Tell them that if we made a waterproof box that just fit around this cube, its capacity would be exactly 1 milliliter. 1 milliliter of water fills the same amount of space as 1 cubic centimeter.	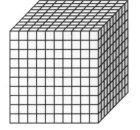
Show students the 1000-cube from a base-10 set. Ask them how many milliliters of water a container that fits exactly around the cubes would hold. Since 1 ml is the same as 1 cm^3, and there are 1000 cm^3 in a block, it would hold 1000 ml. Compare the cube to the liter of water in the measuring cup. Tell them that if we could pour a liter into the cube, it would fill it up completely. The volume of the cube is the same as 1 liter. 1 liter is 1000 cm^3.	1000 cm^3 = 1 ℓ 1000 cm^3 is NOT 1 m^3
Ask students if a liter is the same as 1 m^3. If they did the enrichment activities from the previous lessons, they would know it is not. If they did not do them, tell them that a cubic meter is 100 cm on the side (not 10, as in the thousand-cube), so its volume is 100 cm x 100 cm x 100 cm = 1,000,000 cm^3. A cubic meter is a million cubic centimeters, or a thousand liters. We cannot say that 1000 cm^3 = 1 m^3 or that 1 m^3 = 1 liter.	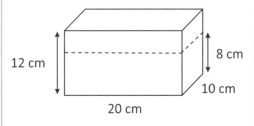
Draw and label a box 20 cm by 10 cm by 12 cm. Ask students to find the volume in cubic centimeters, in milliliters, and in liters and milliliters. (Note: 2000 cm^3 is not the same as 2 m^3. Students should not try to convert 2400 cm^3 first to 2 m^3 400 cm^3. They need to convert all the cubic centimeters to milliliters first, then to liters and milliliters. just because 2400 cm is the same as 24 m.)	Volume = 20 cm x 10 cm x 12 cm = 2400 cm^3 = 2400 ml = 2 ℓ 400 ml Volume of water = 20 cm x 10 cm x 8 cm = 1600 ml = 1 ℓ 600 ml

Then tell students that this box is a tank and is filled with water to 8 cm. Ask them for the volume of water in the tank, and how much more water is needed to fill the tank. Discuss two methods to find the answer to the second part; subtract from the total volume or multiply the length, width, and height of the remaining part of the tank.	Additional water needed to fill tank: 2 ℓ 400 ml – 1 ℓ 600 ml = 800 ml or 20 cm x 10 ml x 4 ml = 800 ml
Discussion	**Text p. 97, Tasks 7-10**

Task 7: This task reiterates the concrete introduction.
 	
Tasks 8-9: Since 1 ml = 1 cm^3 and 1 ℓ = 1000 cm^3, the process for converting from liters to cubic centimeters is the same as converting from liters to milliliters and vice versa.
	7. 1000 cm^3
	   1000 cm^3
	   1 cm^3
	
	8. (a) 2000 cm^3 (b) 400 cm^3 (c) 1200 cm^3
	
	9. (a) 1 ℓ 750 ml (b) 2 ℓ 450 ml (c) 3 ℓ 50 ml
	
	10. (a) 12,000 cm^3
	   (b) 4800 cm^3
	     4 ℓ 800 ml

Assessment	**Text p. 99, Practice 6B**

	1. (a) 3000 cm^3 (b) 250 cm^3 (c) 2060 cm^3
	
	2. (a) 1 ℓ 50 ml (b) 1 ℓ 800 ml (c) 3 ℓ 500 ml
	
	3. 15 cm x 10 cm x 3 cm = 450 cm^3 = **450 ml**
	   The tin can hold 450 ml.
	
	4. (a) 18 cm x 20 cm x 8 cm = **2880 cm^3**
	   (b) 2880 cm^3 = 2880 ml = **2 ℓ 880 ml**
	
	5. (a) 25 cm x 30 cm x 20 cm = **15,000 cm^3**
	   (b) 15 cm x 10 cm x 20 cm = **3000 cm^3**

Practice	WB Exercise 51, pp. 115-116

Exercise 51

1. (a) 300 cm^3 (b) 800 cm^3

2. (a) 400 ml   (b) 120 ml

3. (a) 4 ℓ   (b) 3 ℓ

4. 1 ℓ 200 ml   3 ℓ 600 ml

   1 ℓ 200 ml   3 ℓ 600 ml

   2 ℓ 160 ml   1 ℓ 440 ml

115

116

Objectives

♦ Review.

Note

Provide more capable students with more challenging problems from one of the supplements.

This is the last review in the book, and will give you an opportunity to assess how well you taught the material and whether any topic needs some re-teaching.

Review	Text pp. 100-104, Review E
	1. (a) 79,431; 79,433; 80,331; 80,431 (b) 0.09, 0.55, 0.6, 0.7 (c) $2\frac{2}{9}, 2\frac{4}{9}, 2\frac{2}{3}, \frac{9}{2}$ 2. (a) $1\frac{1}{4}$ (b) $1\frac{2}{9}$ (c) $4\frac{7}{10}$ (d) $\frac{1}{6}$ (e) 8 (f) $13\frac{1}{3}$ 3. Length of flower bed: 25 m – 14 m = 11 m Width of flower bed: 20 m – 14 m = 6 m Area of flower bed: 11 m x 6 m = **66 m²** 4. $\frac{3}{5}$ x 600 g = 3 x 120 g = **360 g** She used 360 g of flour. 5. 6 x 2 x 3 = **36** 36 cubes are needed. 6. (a) 2 units = $18 1 unit = $18 ÷ 2 = $9 3 units = $9 x 3 = **$27** The total amount of money is $27. (b) 8

7. (a) 1.6   (b) $2\frac{1}{20}$

8. (a) 22,548   (b) 38,412   (c) 14,455

9. 4

10. (a) 6
    (b) 2

11.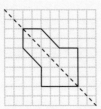

12.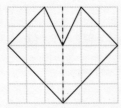

13.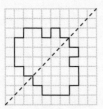

14.

15. (a) Perimeter: 110 cm
       Area: 460 cm^2
    (b) Perimeter: 60 m
       Area: 128 m^2

16. (a) Area of outside rectangle: 21 m x 11 m = 231 m^2
       Area of inside rectangle: 15 m x 5 m = 75 m^2
       Area of shaded part: 231 m^2 – 75 m^2 = **156 m^2**
    (b) Area of large rectangle: 13 m x 11 m = 143 m^2
       Area of small rectangle: 5 m x 7 m = 35 m^2
       Area of shaded part: 143 m^2 – 35 m^2 = **108 m^2**

17. Length: 78 m^2 ÷ 6 m = **13 m**
    Perimeter: 13 m + 6 m + 13 m + 6 m = **38 m**

18. Side: 48 in. ÷ 4 = 12 in.
    Area: 12 in. x 12 in. = **144 in.^{2}**

19. 4.5 m ÷ 5 = **0.9 m** or **90 cm**
    She used 0.9 m for each pillowcase.

20. 5 weeks: $70.50
    1 week: $70.50 ÷ 5 = $14.10
    8 weeks: $14.10 x 8 = **$112.80**
    She would save $112.80 in 8 weeks.

21.

    2 units = 200 – 40 = 160
    1 unit = 160 ÷ 2 = **80**
    80 tickets were sold on Monday.

22. (a) $8
    (b) March
    (c) $64

6.2d   Review (continued)

Objectives

♦ Review.

Review	WB Reviews 6-7, pp. 117-128

Review 6

1. (a) 79,031 (b) 55,100
   (c) 23.29 (d) 18.21
2. (a) 1   (b) 9
3. (a) $35,500
   (b) 8 m
   (c) 17 yd
4. (a) $\frac{2}{3}$; 1
   (b) 3.15, 3.35
5. 3.05
6. 0.4

117

7. $\frac{5}{7}$

8. $2\frac{2}{5}$

9. (a) 6.66   (b) 0.27
   (c) 24   (d) 0.55

10. **9**   (3 thirds in 1 so
       3 x 3 = 9 thirds in 3.)

11. 4.3
12. 1 kg 585 g
13. $2250 (See solution below.)
14. $64.85

118

15. $\frac{32}{40} = \frac{4}{5}$

16. $\frac{1}{10}$ of $20 = $20 ÷ 10 = **$2**

17. Answers will vary.

18.

125°

19. 4 cm^3

119

13. 4 units = $1800
    1 unit = $1800 ÷ 4 = $450
    5 units = $450 x 5 = **$2250**

$1800
TV
Camera
?

20.

Reis
Ben $14.80
Sam
$61.20
1 unit = $61.20 – $14.80
       = $46.40
2 units = $46.40 x 2 = **$92.80**
Reis received $92.80.

21. width: 54 cm^2 ÷ 9 cm = 6 cm
    length + width:
    6 cm + 9 cm = 15 cm
    perimeter: 15 cm x 2 = **30 cm**
    The perimeter is 30 cm.

120

22. Amount left after dress:
    4.5 ft – 0.9 ft = 3.6 ft
    Amount for each cushion:
    3.6 ft ÷ 5 = **0.72 ft**
    She used 0.72 ft for each
    cushion.

23.

? $14
1 unit = $14 ÷ 2 = $7
3 units = $7 x 3 = **$21**
The racket cost $21.

121

Review 7

1. (a) 1000   (b) 0.1
2. 40.6
3. 6
4. (a) 1, 2, 4, 5, 10, 20
   (b) 1, 2, 4
5. 3500 + 500 + 9600 = **13,600**
6. 147.3 lb
7. 4.32
8. $\frac{3}{8}$
9. 1.5

122

---

10. (a) 1 ℓ 540 ml
    (b) 3650 m
11. 1 ℓ 200ml ÷ 3 = **400 ml**
12. 3.82 m x 6 = **22.92 m**
13. 8 ft ÷ 6 = **1.3 ft**
14. $1.50 x 5 = $7.50
    $7.50 + $4.50 = **$12**
15. There are three 50 cents in $1.50, 1 was used, so $\frac{2}{3}$ are left.
16. $\frac{2}{5}$ of $840 = 2 x $168 = **$336**

123

---

17. 45 yd + 20 yd = 65 yd
    65 yd x 2 = 130 yd
    130 yd x 5 = **650 yd**
18. 2
19. 9 cm^3
20. 15 m x 6 m x 4 m = **360 m^3**

124

---

21. Width: 30 cm + 6 cm = 36 cm
    Length: 24 cm + 6 cm = 30 cm
    36 cm + 30 cm = 66
    66 cm x 2 = **132 cm**
    The perimeter of the card is 132 cm.

22.
    3 units = $45
    1 unit = $45 ÷ 3 = **$15**
    He had $15 left.

125

---

23. (a) 6 + 14 + 10 + 4 = **34**
    34 students wear glasses.

    (b)
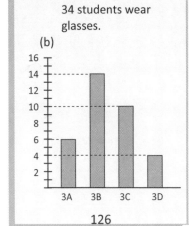

126

---

24. $\frac{1}{7}$ x 30.1 gal = **4.3 gal**

    The bucket has a capacity of 4.3 gal.

25. Perimeter = 6 units = 30 in.
    Width = 1 unit
    = 30 ÷ 6 = 5 in.
    Length = 2 units
    = 5 in. x 2 = 10 in.
    Area = 5 in. x 10 in. = **50 in.^{2}**
    The area of the rectangle is 50 in.^{2}

127

---

26. 15 quarters
    = $0.25 x 15 = $3.75
    35 nickels
    = $0.05 x 35 = $1.75
    21 dimes
    = $0.10 x 21 = $2.10
    $3.75 + $1.75 + $2.10 = **$7.60**
    He saved $7.60 in the three months.

27. $116 x 4 = 464 quarters
    $0.75 has 3 quarters
    464 + 3 = **467**
    There are 467 quarters.

128

Answers to Mental Math

Mental Math 1	Mental Math 2	Mental Math 3	Mental Math 4	Mental Math 5
Write the decimal.	25 tenths = **2.5**	Write the decimal.	7 hundredths = **0.07**	0.34 + 0.03 = **0.37**
$\frac{1}{10}$ = **0.1**	11 tenths = **1.1**	$\frac{1}{2}$ = **0.5**	17 hundredths = **0.17**	0.89 − 0.03 = **0.86**
$\frac{3}{5}$ = **0.6**	87 tenths = **8.7**	$\frac{1}{50}$ = **0.02**	170 hundredths = **1.7**	0.45 + 0.4 = **0.85**
$\frac{1}{2}$ = **0.5**	10 tenths = **1**	$\frac{3}{5}$ = **0.6**	1700 hundredths = **17**	0.68 − 0.5 = **0.18**
$\frac{1}{5}$ = **0.2**	100 tenths = **10**	$\frac{1}{20}$ = **0.05**	4 tenths 5 hundredths = **0.45**	0.38 + 0.3 = **0.68**
$\frac{7}{10}$ = **0.7**	123 tenths = **12.3**	$\frac{1}{4}$ = **0.25**	4 tenths 15 hundredths = **0.55**	0.22 − 0.04 = **0.18**
$\frac{4}{5}$ = **0.8**	842 tenths = **84.2**	$\frac{57}{100}$ = **0.57**	40 tenths 5 hundredths = **4.05**	0.72 + 0.5 = **1.22**
$\frac{3}{10}$ = **0.3**	104 tenths = **10.4**	$\frac{3}{10}$ = **0.3**	14 tenths 5 hundredths = **1.45**	0.18 − 0.09 = **0.09**
$\frac{2}{5}$ = **0.4**	6 ones 14 tenths = **7.4**	$\frac{3}{100}$ = **0.03**	6 hundredths = **0.06**	0.73 + 0.9 = **1.63**
0.4 + **0.6** = 1	11 ones 10 tenths = **12**	$\frac{2}{5}$ = **0.4**	41 hundredths = **0.41**	0.52 + 0.08 = **0.6**
0.5 + **0.5** = 1	72 ones 72 tenths = **79.2**	$\frac{1}{25}$ = **0.04**	8 hundredths = **0.08**	0.93 − 0.2 = **0.73**
0.7 + **0.3** = 1	90 ones 4 tenths = **90.4**	$\frac{3}{4}$ = **0.75**	805 hundredths = **8.05**	0.89 − 0.05 = **0.84**
0.2 + **0.8** = 1	1 + 0.5 = **1.5**	$\frac{13}{100}$ = **0.13**	5 + 0.8 + 0.09 = **5.89**	0.66 − 0.04 = **0.62**
0.1 + **0.9** = 1	30 + 5 + 0.7 = **35.7**	$2\frac{1}{25}$ = **2.04**	45 + 0.65 = **45.65**	0.92 − 0.05 = **0.87**
0.3 + **0.7** = 1	10 + 4 + 0.8 = **14.8**	$1\frac{1}{20}$ = **1.05**	0.03 + 1 + 0.7 = **1.73**	0.22 + 0.05 = **0.27**
0.9 + **0.1** = 1	9 + 10 + 0.2 = **19.2**	$\frac{3}{2}$ = **1.5**	0.9 + 0.08 + 50 = **50.98**	0.47 + 0.7 = **1.17**
0.6 + **0.4** = 1	0.9 + 2 + 50 = **52.9**		100 + 0.01 = **100.01**	0.17 + 0.08 = **0.25**
0.8 + **0.2** = 1	0.5 + 80 = **80.5**		0.02 + 14.9 = **14.92**	0.88 − 0.4 = **0.48**
	40 + 0.3 + 7 = **47.3**		0.7 + 23.02 = **23.72**	0.9 + 0.06 = **0.96**
	0.1 + 100 = **100.1**		103.7 + 50 + 0.05 = **153.75**	0.6 − 0.03 = **0.57**

Mental Math 6	Mental Math 7	Mental Math 8	Mental Math 9	Mental Math 10	Mental Math 11
3.67 + 0.4 = **4.07**	0.003 + 0.002 = **0.005**	6.002 + 0.05 = **6.052**	5.1 + 0.9 = **6**	0.72 + 0.06 = **0.78**	0.52 + 0.73 = **1.25**
6.88 − 0.06 = **6.82**	0.014 − 0.003 = **0.011**	1.506 − 0.3 = **1.206**	8.8 + 0.2 = **9**	0.48 + 0.6 = **1.08**	0.48 + 0.34 = **0.82**
2.32 − 0.7 = **1.62**	0.126 + 0.01 = **0.136**	3.896 + 0.002 = **3.898**	9.3 + 0.2 = **9.5**	0.09 + 0.59 = **0.68**	0.32 + 0.78 = **1.1**
1.6 − 0.08 = **1.52**	0.209 + 0.8 = **1.009**	0.119 + 0.03 = **0.149**	7.9 + 0.6 = **8.5**	0.63 + 0.5 = **1.13**	0.67 + 0.43 = **1.1**
6.36 − 0.08 = **6.28**	2.231 − 0.007 = **2.224**	1.648 − 0.2 = **1.448**	8.2 + 4.1 = **12.3**	0.16 + 0.04 = **0.2**	0.91 + 0.49 = **1.4**
1.64 + 0.07 = **1.71**	5.198 − 0.008 = **5.19**	7.12 − 0.009 = **7.111**	4.3 + 5.7 = **10**	0.62 + 0.8 = **1.42**	0.95 + 0.35 = **1.3**
8.4 − 0.06 = **8.34**	6.218 + 0.05 = **6.268**	4.028 + 0.002 = **4.03**	6.6 + 1.5 = **8.1**	0.92 + 0.08 = **1**	0.29 + 0.98 = **1.27**
2.28 + 0.8 = **3.08**	0.171 − 0.06 = **0.111**	8.1 + 0.005 = **8.105**	6.6 + 2.6 = **9.2**	0.42 + 0.8 = **1.22**	0.59 + 0.62 = **1.21**
3.7 − 0.04 = **3.66**	9.01 + 0.009 = **9.019**	6.991 − 0.4 = **6.591**	8 + 4.2 = **12.2**	0.91 + 0.03 = **0.94**	0.24 + 0.87 = **1.11**
5.19 + 0.01 = **5.2**	8.842 + 0.6 = **9.442**	0.4 − 0.004 = **0.396**	9.5 + 1.3 = **10.8**	0.58 + 0.6 = **1.18**	0.36 + 0.77 = **1.13**
7.48 − 0.5 = **6.98**	0.142 − 0.04 = **0.102**	1.875 − 0.003 = **1.872**	5.2 + 0.7 = **5.9**	0.62 + 0.09 = **0.71**	0.42 + 0.99 = **1.41**
6.44 + 0.07 = **6.51**	7.541 − 0.003 = **7.538**	4.172 − 0.8 = **3.372**	0.8 + 3.4 = **4.2**	0.86 + 0.04 = **0.9**	0.28 + 0.44 = **0.72**
1.32 + 0.8 = **2.12**	3.9 + 0.08 = **3.98**	7.052 + 0.007 = **7.059**	3.2 + 7.8 = **11**	0.65 + 0.8 = **1.45**	0.92 + 0.95 = **1.87**
9.99 + 0.9 = **10.89**	1.472 + 0.8 = **2.272**	9.204 − 0.2 = **9.004**	6.7 + 4.5 = **11.2**	0.34 + 0.06 = **0.4**	0.42 + 0.97 = **1.39**
9.99 + 0.09 = **10.08**	2.355 + 0.005 = **2.36**	2.632 − 0.06 = **2.572**	9.1 + 4.9 = **14**	0.03 + 0.69 = **0.72**	0.82 + 0.71 = **1.53**
1.5 − 0.03 = **1.47**	9.105 − 0.5 = **8.605**	3.311 + 0.7 = **4.011**	9.5 + 3.5 = **13**	0.41 + 0.9 = **1.31**	0.48 + 0.65 = **1.13**
5 − 0.3 = **4.7**	3.4 + 0.03 = **3.43**	0.985 + 0.006 = **0.991**	2.9 + 9.8 = **12.7**	0.78 + 0.07 = **0.85**	0.34 + 0.55 = **0.89**
5 − 0.03 = **4.97**	3.4 − 0.03 = **3.37**	6.324 − 0.09 = **6.234**	5.9 + 6.2 = **12.1**	0.49 + 0.02 = **0.51**	0.68 + 0.84 = **1.52**
10 − 0.7 = **9.3**	3.4 − 0.003 = **3.397**	9.999 + 0.001 = **10**	2.4 + 8.7 = **11.1**	4.48 + 0.9 = **5.38**	0.99 + 0.44 = **1.43**
10 − 0.07 = **9.93**	10 − 0.005 = **9.995**	9.999 + 0.01 = **10.009**	3.6 + 7.7 = **11.3**	3.27 + 0.7 = **3.97**	0.77 + 0.82 = **1.59**

Answers to Mental Math

Mental Math 12	Mental Math 13	Mental Math 14	Mental Math 15	Mental Math 16	Mental Math 17
4.9 – 0.5 = **4.4**	0.85 – 0.02 = **0.83**	0.98 – 0.03 = **0.95**	6.9 – 2.8 = **4.1**	5.54 – 0.98 = **4.56**	0.4 x 8 = **3.2**
9.6 – 0.3 = **9.3**	5.69 – 0.08 = **5.61**	1.84 – 0.04 = **1.8**	5.6 – 3.2 = **2.4**	7.22 – 1.99 = **5.23**	0.7 x 7 = **4.9**
4.2 – 0.8 = **3.4**	0.1 – 0.08 = **0.02**	4.73 – 0.09 = **4.64**	5.6 – 1.8 = **3.8**	3.83 – 2.95 = **0.88**	0.2 x 9 = **1.8**
3.3 – 0.7 = **2.6**	0.9 – 0.04 = **0.86**	3.5 – 0.02 = **3.48**	9.7 – 6.4 = **3.3**	7.47 – 4.96 = **2.51**	0.06 x 2 = **0.12**
2.5 – 0.6 = **1.9**	9.6 – 0.08 = **9.52**	2.3 – 0.07 = **2.23**	9.5 – 3.8 = **5.7**	8.09 – 3.98 = **4.11**	0.03 x 8 = **0.24**
8.1 – 0.9 = **7.2**	6.5 – 0.07 = **6.43**	4.66 – 0.09 = **4.57**	5.2 – 4.8 = **0.4**	6.66 + 2.95 = **9.61**	7 x 0.5 = **3.5**
7.3 – 0.6 = **6.7**	4.3 – 0.02 = **4.28**	3.53 – 0.05 = **3.48**	8.4 – 4.5 = **3.9**	3.97 + 2.22 = **6.19**	6 x 0.06 = **0.36**
5.4 – 0.8 = **4.6**	3.95 – 0.03 = **3.92**	2.6 – 0.06 = **2.54**	6 – 3.7 = **2.3**	8.71 – 2.96 = **5.75**	0.09 x 8 = **0.72**
3.7 – 0.2 = **3.5**	1 – 0.07 = **0.93**	2.42 – 0.08 = **2.34**	4.1 – 2.2 = **1.9**	6.02 – 4.98 = **1.04**	0.7 x 8 = **5.6**
6.5 – 0.2 = **6.3**	2 – 0.09 = **1.91**	4.55 – 0.09 = **4.46**	5.8 – 2.3 = **3.5**	3.87 + 3.95 = **7.82**	0.3 x 9 = **2.7**
6.3 – 0.7 = **5.6**	8 – 0.06 = **7.94**	2.6 – 0.04 = **2.56**	8 – 6.5 = **1.5**	7.01 – 2.98 = **4.03**	0.06 x 4 = **0.24**
8.2 – 0.3 = **7.9**	1 – 0.38 = **0.62**	6.92 – 0.09 = **6.83**	4.2 – 2.4 = **1.8**	5.57 – 3.97 = **1.6**	4 x 0.05 = **0.2**
9.4 – 0.6 = **8.8**	1 – 0.76 = **0.24**	1.22 – 0.05 = **1.17**	6.4 – 2.5 = **3.9**	1.99 + 6.23 = **8.22**	6 x 0.9 = **5.4**
8.5 – 0.5 = **8**	1 – 0.33 = **0.67**	4.85 – 0.06 = **4.79**	5.5 – 2.8 = **2.7**	4.03 – 2.96 = **1.07**	0.8 x 2 = **1.6**
7.9 – 0.4 = **7.5**	3 – 0.49 = **2.51**	2.32 – 0.06 = **2.26**	8.4 – 6.1 = **2.3**	5.2 – 3.99 = **1.21**	0.03 x 6 = **0.18**
6.21 – 0.9 = **5.31**	5 – 0.75 = **4.25**	2.83 – 0.08 = **2.75**	5.3 – 3.8 = **1.5**	6.4 – 2.98 = **3.42**	0.5 x 3 = **1.5**
4.36 – 0.9 = **3.46**	6 – 0.84 = **5.16**	0.36 – 0.08 = **0.28**	7 – 3.6 = **3.4**	3.7 + 0.98 = **4.68**	0.8 x 5 = **4**
8.04 – 0.6 = **7.44**	8 – 0.43 = **7.57**	3.74 – 0.07 = **3.67**	7.7 – 3.8 = **3.9**	4.8 + 4.97 = **9.77**	0.06 x 7 = **0.42**
3.24 – 0.7 = **2.54**	3 – 0.85 = **2.15**	2.87 – 0.08 = **2.79**	8.1 – 4.9 = **3.2**	3.05 + 0.98 = **4.03**	9 x 0.09 = **0.81**
2.3 – 0.4 = **1.9**	5 – 0.66 = **4.34**	2.43 – 0.06 = **2.37**	5.3 – 3.5 = **1.8**	6.5 – 0.95 = **5.55**	0.2 x 5 = **1**

Mental Math 18		Mental Math 19	Mental Math 20	Mental Math 21
7.7 x 3 = 21 + **2.1** = **23.1**		1.5 ÷ 5 = **0.3**	6.26 + 0.4 = **6.66**	10 min 5 s – 50 s = **9** min **15** s
3.4 x 5 = **15** + 2 = **17**		6.4 ÷ 8 = **0.8**	0.06 x 5 = **0.3**	15 km 5 m – 50 m = **14** km **955** m
8.3 x 6 = **48** + 1.8 = **49.8**		0.42 ÷ 7 = **0.06**	0.174 – 0.01 = **0.164**	20 m 5 cm – 50 cm = **19** m **55** cm
4.3 x 4 = **16** + 1.2 = **17.2**		0.3 ÷ 6 = **0.05**	7.8 + 2.5 = **10.3**	8 ft 5 in – 10 in. = **7** ft **7** in.
4.5 x 6 = **24** + 3 = **27**		5.4 ÷ 9 = **0.6**	6.36 – 0.08 = **6.28**	13 lb 5 oz – 10 oz = **12** lb **11** oz
8.5 x 5 = **40** + 2.5 = **42.5**		0.24 ÷ 4 = **0.06**	0.63 ÷ 7 = **0.09**	11 min 50 s + 50 s = **12** min **40** s
6.2 x 7 = **42** + 1.4 = **43.4**		0.2 ÷ 5 = **0.04**	0.1 – 0.06 = **0.04**	21 km 600 m + 600 m = **22** km **200** m
3.3 x 5 = **15** + 1.5 = **16.5**		8.1 ÷ 9 = **0.9**	6.3 – 0.98 = **5.32**	5 m 60 cm + 60 cm = **6** m **20** cm
0.43 x 3 = **1.2** + **0.09** = **1.29**		0.49 ÷ 7 = **0.07**	9.4 – 0.3 = **9.1**	3 ft 10 in + 10 in. = **4** ft **8** in.
0.63 x 8 = **4.8** + **0.24** = **5.04**		4 ÷ 8 = **0.5**	5 – 0.04 = **4.96**	13 lb 10 oz + 10 oz = **14** lb **4** oz
		0.27 ÷ 9 = **0.03**	0.04 x 8 = **0.32**	10 min 10 s x 10 = **101** min **40** s
2.3 x 2 = **4.6**	2.8 x 3 = **8.4**	0.21 ÷ 3 = **0.07**	6.73 – 0.08 = **6.65**	10 ℓ 10 ml x 10 = **100** ℓ **100** ml
5.6 x 3 = **16.8**	8.2 x 7 = **57.4**	0.25 ÷ 5 = **0.05**	4.56 + 0.99 = **5.55**	10 m 10 cm x 10 = **101** m **0** cm
3.6 x 7 = **25.2**	4.9 x 4 = **19.6**	4.5 ÷ 5 = **0.9**	2.6 – 0.07 = **2.53**	10 ft 10 in x 10 = **108** ft **4** in.
7.1 x 2 = **14.2**	3.6 x 2 = **7.2**	0.16 ÷ 2 = **0.08**	7.21 – 3.97 = **3.24**	10 lb 10 oz x 10 = **106** lb **4** oz
2.1 x 4 = **8.4**	0.14 x 2 = **0.28**	3.2 ÷ 4 = **0.8**	0.56 + 0.08 = **0.64**	10 gal 3 qt x 10 = **107** gal **2** qt
6.4 x 5 = **32**	0.26 x 3 = **0.78**	3.5 ÷ 5 = **0.7**	2.6 – 0.4 = **2.2**	10 yr 3 months x 10 = **102** yr **6** months
4.7 x 3 = **14.1**	0.83 x 3 = **2.49**	2.4 ÷ 8 = **0.3**	6.218 + 0.1 = **6.318**	10 weeks 3 days x 10 = **104** weeks **2** days
3.1 x 4 = **12.4**	0.27 x 5 = **1.35**	0.48 ÷ 6 = **0.08**	0.3 ÷ 5 = **0.06**	10 yd 1 ft x 10 = **103** yd **1** ft
6.9 x 8 = **55.2**	0.55 x 4 = **2.2**	1.8 ÷ 6 = **0.3**	1.5 – 0.8 = **0.7**	2 qt 1 pt x 10 = **25** qt **0** pt

Blank Page

Appendix

Mental Math 1	Mental Math 2	Mental Math 3
Write the decimal.	25 tenths = _____	Write the decimal
$\frac{1}{10}$ = _____	11 tenths = _____	$\frac{1}{2}$ = _____
$\frac{3}{5}$ = _____	87 tenths = _____	$\frac{1}{50}$ = _____
$\frac{1}{2}$ = _____	10 tenths = _____	$\frac{3}{5}$ = _____
$\frac{1}{5}$ = _____	100 tenths = _____	$\frac{1}{20}$ = _____
$\frac{7}{10}$ = _____	123 tenths = _____	$\frac{1}{4}$ = _____
$\frac{4}{5}$ = _____	842 tenths = _____	$\frac{57}{100}$ = _____
$\frac{3}{10}$ = _____	104 tenths = _____	$\frac{3}{10}$ = _____
$\frac{2}{5}$ = _____	6 ones 14 tenths = _____	$\frac{3}{100}$ = _____
0.4 + _____ = 1	11 ones 10 tenths = _____	$\frac{2}{5}$ = _____
0.5 + _____ = 1	72 ones 72 tenths = _____	$\frac{1}{25}$ = _____
0.7 + _____ = 1	90 ones 4 tenths = _____	$\frac{3}{4}$ = _____
0.2 + _____ = 1	1 + 0.5 = _____	$\frac{13}{100}$ = _____
0.1 + _____ = 1	30 + 5 + 0.7 = _____	$2\frac{1}{25}$ = _____
0.3 + _____ = 1	10 + 4 + 0.8 = _____	$1\frac{1}{20}$ = _____
0.9 + _____ = 1	9 + 10 + 0.2 = _____	$\frac{3}{2}$ = _____
0.6 + _____ = 1	0.9 + 2 + 50 = _____	
0.8 + _____ = 1	0.5 + 80 = _____	
	40 + 0.3 + 7 = _____	
	0.1 + 100 = _____	

Mental Math 4	Mental Math 5
7 hundredths = _____	0.34 + 0.03 = _____
17 hundredths = _____	0.89 − 0.03 = _____
170 hundredths = _____	0.45 + 0.4 = _____
1700 hundredths = _____	0.68 − 0.5 = _____
4 tenths 5 hundredths = _____	0.38 + 0.3 = _____
4 tenths 15 hundredths = _____	0.22 − 0.04 = _____
40 tenths 5 hundredths = _____	0.72 + 0.5 = _____
14 tenths 5 hundredths = _____	0.18 − 0.09 = _____
6 hundredths = _____	0.73 + 0.9 = _____
41 hundredths = _____	0.52 + 0.08 = _____
8 hundredths = _____	0.93 − 0.2 = _____
805 hundredths = _____	0.89 − 0.05 = _____
5 + 0.8 + 0.09 = _____	0.66 − 0.04 = _____
45 + 0.65 = _____	0.92 − 0.05 = _____
0.03 + 1 + 0.7 = _____	0.22 + 0.05 = _____
0.9 + 0.08 + 50 = _____	0.47 + 0.7 = _____
100 + 0.01 = _____	0.17 + 0.08 = _____
0.02 + 14.9 = _____	0.88 − 0.4 = _____
0.7 + 23.02 = _____	0.9 + 0.06 = _____
103.7 + 50 + 0.05 = _____	0.6 − 0.03 = _____

Mental Math 6	Mental Math 7	Mental Math 8
3.67 + 0.4 = _____	0.003 + 0.002 = _____	6.002 + 0.05 = _____
6.88 − 0.06 = _____	0.014 − 0.003 = _____	1.506 − 0.3 = _____
2.32 − 0.7 = _____	0.126 + 0.01 = _____	3.896 + 0.002 = _____
1.6 − 0.08 = _____	0.209 + 0.8 = _____	0.119 + 0.03 = _____
6.36 − 0.08 = _____	2.231 − 0.007 = _____	1.648 − 0.2 = _____
1.64 + 0.07 = _____	5.198 − 0.008 = _____	7.12 − 0.009 = _____
8.4 − 0.06 = _____	6.218 + 0.05 = _____	4.028 + 0.002 = _____
2.28 + 0.8 = _____	0.171 − 0.06 = _____	8.1 + 0.005 = _____
3.7 − 0.04 = _____	9.01 + 0.009 = _____	6.991 − 0.4 = _____
5.19 + 0.01 = _____	8.842 + 0.6 = _____	0.4 − 0.004 = _____
7.48 − 0.5 = _____	0.142 − 0.04 = _____	1.875 − 0.003 = _____
6.44 + 0.07 = _____	7.541 − 0.003 = _____	4.172 − 0.8 = _____
1.32 + 0.8 = _____	3.9 + 0.08 = _____	7.052 + 0.007 = _____
9.99 + 0.9 = _____	1.472 + 0.8 = _____	9.204 − 0.2 = _____
9.99 + 0.09 = _____	2.355 + 0.005 = _____	2.632 − 0.06 = _____
1.5 − 0.03 = _____	9.105 − 0.5 = _____	3.311 + 0.7 = _____
5 − 0.3 = _____	3.4 + 0.03 = _____	0.985 + 0.006 = _____
5 − 0.03 = _____	3.4 − 0.03 = _____	6.324 − 0.09 = _____
10 − 0.7 = _____	3.4 − 0.003 = _____	9.999 + 0.001 = _____
10 − 0.07 = _____	10 − 0.005 = _____	9.999 + 0.01 = _____

Mental Math 9	Mental Math 10	Mental Math 11
5.1 + 0.9 = _____	0.72 + 0.06 = _____	0.52 + 0.73 = _____
8.8 + 0.2 = _____	0.48 + 0.6 = _____	0.48 + 0.34 = _____
9.3 + 0.2 = _____	0.09 + 0.59 = _____	0.32 + 0.78 = _____
7.9 + 0.6 = _____	0.63 + 0.5 = _____	0.67 + 0.43 = _____
8.2 + 4.1 = _____	0.16 + 0.04 = _____	0.91 + 0.49 = _____
4.3 + 5.7 = _____	0.62 + 0.8 = _____	0.95 + 0.35 = _____
6.6 + 1.5 = _____	0.92 + 0.08 = _____	0.29 + 0.98 = _____
6.6 + 2.6 = _____	0.42 + 0.8 = _____	0.59 + 0.62 = _____
8 + 4.2 = _____	0.91 + 0.03 = _____	0.24 + 0.87 = _____
9.5 + 1.3 = _____	0.58 + 0.6 = _____	0.36 + 0.77 = _____
5.2 + 0.7 = _____	0.62 + 0.09 = _____	0.42 + 0.99 = _____
0.8 + 3.4 = _____	0.86 + 0.04 = _____	0.28 + 0.44 = _____
3.2 + 7.8 = _____	0.65 + 0.8 = _____	0.92 + 0.95 = _____
6.7 + 4.5 = _____	0.34 + 0.06 = _____	0.42 + 0.97 = _____
9.1 + 4.9 = _____	0.03 + 0.69 = _____	0.82 + 0.71 = _____
9.5 + 3.5 = _____	0.41 + 0.9 = _____	0.48 + 0.65 = _____
2.9 + 9.8 = _____	0.78 + 0.07 = _____	0.34 + 0.55 = _____
5.9 + 6.2 = _____	0.49 + 0.02 = _____	0.68 + 0.84 = _____
2.4 + 8.7 = _____	4.48 + 0.9 = _____	0.99 + 0.44 = _____
3.6 + 7.7 = _____	3.27 + 0.7 = _____	0.77 + 0.82 = _____

Mental Math 12	Mental Math 13	Mental Math 14
4.9 – 0.5 = _____	0.85 – 0.02 = _____	0.98 – 0.03 = _____
9.6 – 0.3 = _____	5.69 – 0.08 = _____	1.84 – 0.04 = _____
4.2 – 0.8 = _____	0.1 – 0.08 = _____	4.73 – 0.09 = _____
3.3 – 0.7 = _____	0.9 – 0.04 = _____	3.5 – 0.02 = _____
2.5 – 0.6 = _____	9.6 – 0.08 = _____	2.3 – 0.07 = _____
8.1 – 0.9 = _____	6.5 – 0.07 = _____	4.66 – 0.09 = _____
7.3 – 0.6 = _____	4.3 – 0.02 = _____	3.53 – 0.05 = _____
5.4 – 0.8 = _____	3.95 – 0.03 = _____	2.6 – 0.06 = _____
3.7 – 0.2 = _____	1 – 0.07 = _____	2.42 – 0.08 = _____
6.5 – 0.2 = _____	2 – 0.09 = _____	4.55 – 0.09 = _____
6.3 – 0.7 = _____	8 – 0.06 = _____	2.6 – 0.04 = _____
8.2 – 0.3 = _____	1 – 0.38 = _____	6.92 – 0.09 = _____
9.4 – 0.6 = _____	1 – 0.76 = _____	1.22 – 0.05 = _____
8.5 – 0.5 = _____	1 – 0.33 = _____	4.85 – 0.06 = _____
7.9 – 0.4 = _____	3 – 0.49 = _____	2.32 – 0.06 = _____
6.21 – 0.9 = _____	5 – 0.75 = _____	2.83 – 0.08 = _____
4.36 – 0.9 = _____	6 – 0.84 = _____	0.36 – 0.08 = _____
8.04 – 0.6 = _____	8 – 0.43 = _____	3.74 – 0.07 = _____
3.24 – 0.7 = _____	3 – 0.85 = _____	2.87 – 0.08 = _____
2.3 – 0.4 = _____	5 – 0.66 = _____	2.43 – 0.06 = _____

Mental Math 15	Mental Math 16	Mental Math 17
6.9 – 2.8 = _____	5.54 – 0.98 = _____	0.4 x 8 = _____
5.6 – 3.2 = _____	7.22 – 1.99 = _____	0.7 x 7 = _____
5.6 – 1.8 = _____	3.83 – 2.95 = _____	0.2 x 9 = _____
9.7 – 6.4 = _____	7.47 – 4.96 = _____	0.06 x 2 = _____
9.5 – 3.8 = _____	8.09 – 3.98 = _____	0.03 x 8 = _____
5.2 – 4.8 = _____	6.66 + 2.95 = _____	7 x 0.5 = _____
8.4 – 4.5 = _____	3.97 + 2.22 = _____	6 x 0.06 = _____
6 – 3.7 = _____	8.71 – 2.96 = _____	0.09 x 8 = _____
4.1 – 2.2 = _____	6.02 – 4.98 = _____	0.7 x 8 = _____
5.8 – 2.3 = _____	3.87 + 3.95 = _____	0.3 x 9 = _____
8 – 6.5 = _____	7.01 – 2.98 = _____	0.06 x 4 = _____
4.2 – 2.4 = _____	5.57 – 3.97 = _____	4 x 0.05 = _____
6.4 – 2.5 = _____	1.99 + 6.23 = _____	6 x 0.9 = _____
5.5 – 2.8 = _____	4.03 – 2.96 = _____	0.8 x 2 = _____
8.4 – 6.1 = _____	5.2 – 3.99 = _____	0.03 x 6 = _____
5.3 – 3.8 = _____	6.4 – 2.98 = _____	0.5 x 3 = _____
7 – 3.6 = _____	3.7 + 0.98 = _____	0.8 x 5 = _____
7.7 – 3.8 = _____	4.8 + 4.97 = _____	0.06 x 7 = _____
8.1 – 4.9 = _____	3.05 + 0.98 = _____	9 x 0.09 = _____
5.3 – 3.5 = _____	6.5 – 0.95 = _____	0.2 x 5 = _____

Mental Math 18	Mental Math 19

Mental Math 18

7.7 x 3 = 21 + 2.1 = _____

3.4 x 5 = _____ + _____ = _____

8.3 x 6 = _____ + _____ = _____

4.3 x 4 = _____ + _____ = _____

4.5 x 6 = _____ + _____ = _____

8.5 x 5 = _____ + _____ = _____

6.2 x 7 = _____ + _____ = _____

3.3 x 5 = _____ + _____ = _____

0.43 x 3 = _____ + _____ = _____

0.63 x 8 = _____ + _____ = _____

2.3 x 2 = _____   2.8 x 3 = _____

5.6 x 3 = _____   8.2 x 7 = _____

3.6 x 7 = _____   4.9 x 4 = _____

7.1 x 2 = _____   3.6 x 2 = _____

2.1 x 4 = _____   0.14 x 2 = _____

6.4 x 5 = _____   0.26 x 3 = _____

4.7 x 3 = _____   0.83 x 3 = _____

3.1 x 4 = _____   0.27 x 5 = _____

6.9 x 8 = _____   0.55 x 4 = _____

Mental Math 19

1.5 ÷ 5 = _____

6.4 ÷ 8 = _____

0.42 ÷ 7 = _____

0.3 ÷ 6 = _____

5.4 ÷ 9 = _____

0.24 ÷ 4 = _____

0.2 ÷ 5 = _____

8.1 ÷ 9 = _____

0.49 ÷ 7 = _____

4 ÷ 8 = _____

0.27 ÷ 9 = _____

0.21 ÷ 3 = _____

0.25 ÷ 5 = _____

4.5 ÷ 5 = _____

0.16 ÷ 2 = _____

3.2 ÷ 4 = _____

3.5 ÷ 5 = _____

2.4 ÷ 8 = _____

0.48 ÷ 6 = _____

1.8 ÷ 6 = _____

Mental Math 20	Mental Math 21
6.26 + 0.4 = _____	10 min 5 s – 50 s = _____ min _____ s
0.06 x 5 = _____	15 km 5 m – 50 m = _____ km _____ m
0.174 – 0.01 = _____	20 m 5 cm – 50 cm = _____ m _____ cm
7.8 + 2.5 = _____	8 ft 5 in – 10 in. = _____ ft _____ in.
6.36 – 0.08 = _____	13 lb 5 oz – 10 oz = _____ lb _____ oz
0.63 ÷ 7 = _____	11 min 50 s + 50 s = _____ min _____ s
0.1 – 0.06 = _____	21 km 600 m + 600 m = _____ km _____ m
6.3 – 0.98 = _____	5 m 60 cm + 60 cm = _____ m _____ cm
9.4 – 0.3 = _____	3 ft 10 in + 10 in. = _____ ft _____ in.
5 – 0.04 = _____	13 lb 10 oz + 10 oz = _____ lb _____ oz
0.04 x 8 = _____	10 min 10 s x 10 = _____ min _____ s
6.73 – 0.08 = _____	10 ℓ 10 ml x 10 = _____ ℓ _____ m
4.56 + 0.99 = _____	10 m 10 cm x 10 = _____ m _____ cm
2.6 – 0.07 = _____	10 ft 10 in x 10 = _____ ft _____ in.
7.21 – 3.97 = _____	10 lb 10 oz x 10 = _____ lb _____ oz
0.56 + 0.08 = _____	10 gal 3 qt x 10 = _____ gal _____ qt
2.6 – 0.4 = _____	10 yr 3 months x 10 = _____ yr _____ months
6.218 + 0.1 = _____	10 weeks 3 days x 10 = _____ weeks _____ days
0.3 ÷ 5 = _____	10 yd 1 ft x 10 = _____ yd _____ ft
1.5 – 0.8 = _____	2 qt 1 pt x 10 = _____ qt _____ pt

Fraction Squares for Tenths

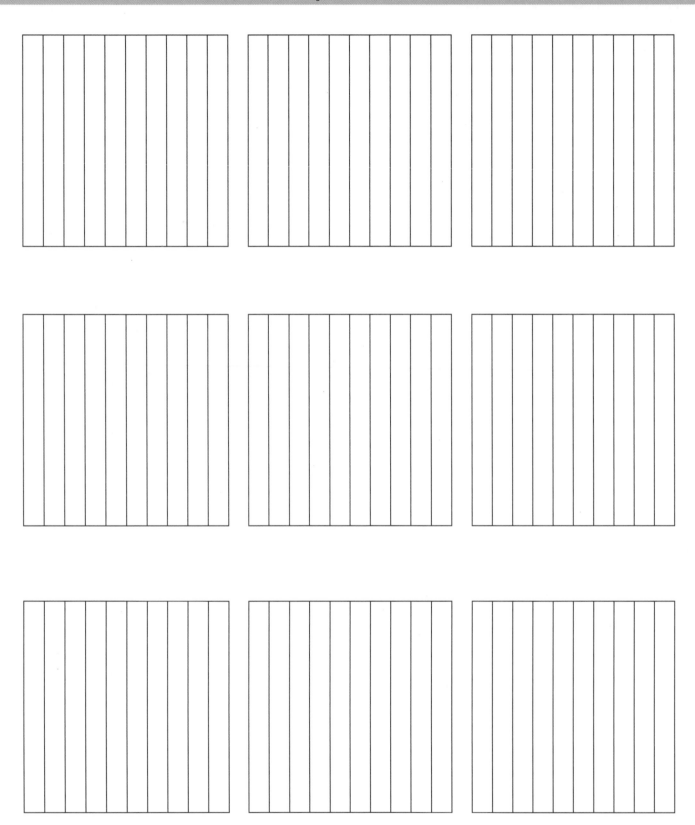

Fraction Squares

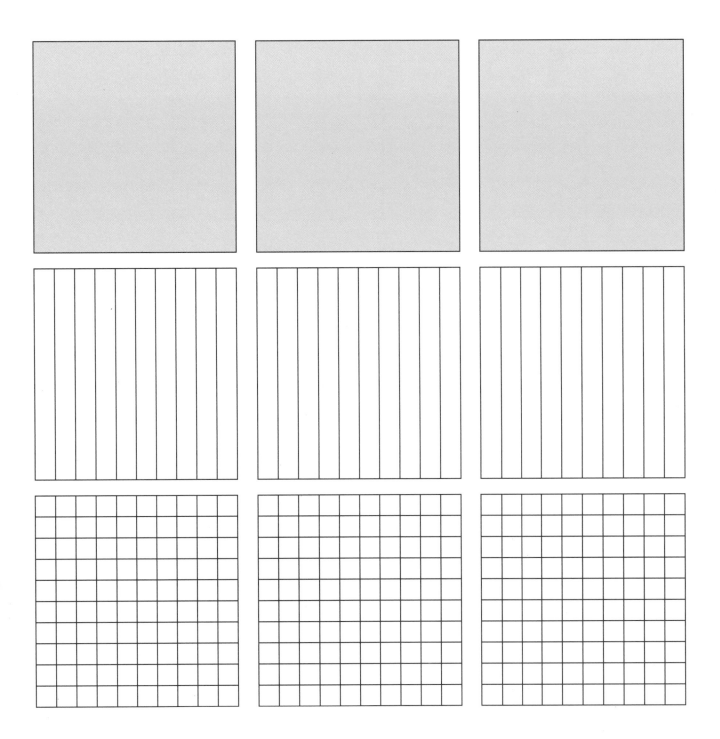

Number Lines - Tenths

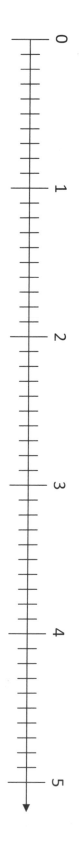

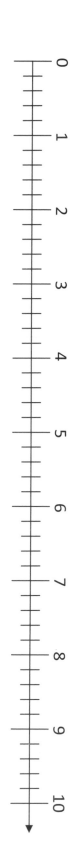

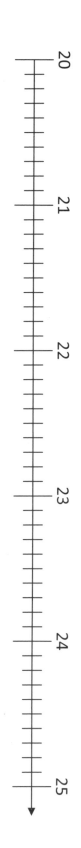

Number Lines - Hundredths

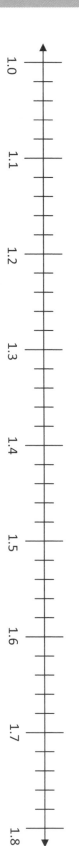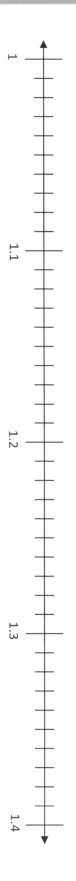

Convert Measures

1. 1 ft = _____ in.

   4 ft = 4 x _____ in. = _____ in.

2. 1 m = _____ cm

   9 m = 9 x _____ cm = _____ cm

3. 1 day = _____ hours

   8 days = 8 x _____ hours = _____ hours

4. 1 lb = _____ oz

   12 lb = 12 x _____ oz = _____ oz

5. 4 ℓ 250 ml

   = (4 x _____) ml + 250 ml

   = _____ ml

6. 5 km 40 m

   = (5 x _____) m + _____ m

   = _____ m

7. 4 years 5 months

   = (4 x _____) months + 5 months

   = _____ months + 5 months

   = _____ months

8. 4 hours 20 minutes

   = (4 x _____) minutes + 20 minutes

   = _____ minutes + 20 minutes

   = _____

9. _____ in. = 1 ft

   48 in. = 48 ÷ 12 = _____ ft

   50 in. = _____ ft _____ in.
   $\wedge$
   48 2   58 ÷ 12 = _____ R _____

10. _____ ft = 1 yd

    8 ft = _____ yd _____ ft
    $\wedge$
    6 2

11. _____ cm = 1 m

    602 cm = _____ m _____ cm
    / \
    600 ___

12. _____ g = 1 kg

    2400 g = _____ kg _____ g
    / \
    ____ 400

13. _____ qt = 1 gal

    365 ÷ 4 = _____ R _____

    365 qt = _____ gal _____ qt

14. _____ h = 1 day

    365 ÷ 24 = _____ R ___

    365 h = _____ days _____ h

Convert Measures - Practice

Fill in the blanks.

1. 20 km = _____ m

2. 84 in. = _____ ft

3. 24 gal = _____ qt

4. 3000 cm = _____ m

5. 100 ℓ = _____ ml

6. 12 years = _____ months

7. 9 yd = _____ in.

8. 4 gal = _____ c

9. 3 lb 3 oz = _____ oz

10. 8 kg 5 g = _____ g

11. 3 qt 1 pt = _____ pt

12. 2 min 24 s = _____ s

13. 20 m 20 cm = _____ cm

14. 20 ℓ 20 ml = _____ ml

15. 200 min = _____ h _____ min

16. 1001 cm = _____ m _____ cm

17. 30 months = _____ years _____ months

18. 100 ft = _____ yd _____ ft

Convert Measures - Answers

Fill in the blanks.

1. 20 km = **20,000** m

2. 84 in. = **7** ft

3. 24 gal = **6** qt

4. 3000 cm = **30** m

5. 100 ℓ = **100,000** ml

6. 12 years = **144** months

7. 9 yd = **324** in.   (9 x 3 ft = 27 ft; 27 ft x 12 in. = 324 in.)

8. 4 gal = **64** c   (4 x 4 qt = 16 qt; 16 x 4 c = 64 c)

9. 3 lb 3 oz = **51** oz

10. 8 kg 5 g = **8005** g

11. 3 qt 1 pt = **7** pt

12. 2 min 24 s = **144** s

13. 20 m 20 cm = **2020** cm

14. 20 ℓ 20 ml = **20,020** ml

15. 200 min = **3** h **20** min

16. 1001 cm = **10** m **1** cm

17. 30 months = **2** years **6** months

18. 100 ft = **33** yd **1** ft

Add and Subtract Measures

1 m – 42 cm = _____ cm

1 km – 390 m = _____ m

1 ℓ – 7 ml = _____ ml

1 ft – 7 in. = _____ in.

1 yd – 2 ft = _____ ft

1 lb – 7 oz = _____ oz

1 h – 15 min = _____ min

4 m – 25 cm = _____ m _____ cm

10 km – 90 m = _____ km _____ m

3 ℓ – 985 ml = _____ ℓ _____ ml

19 ft – 3 in. = _____ ft _____ in.

8 lb – 10 oz = _____ lb _____ oz

5 gal – 3 qt = _____ gal _____ qt

4 h 10 min – 25 min = _____ h _____ min

19 ft 1 in. – 3 in. = _____ ft _____ in.

6 h 10 min – 2 h 25 min = _____ h _____ min

12 m 45 cm – 7 m 80 cm = _____ m _____ cm

65 cm + 40 cm = _____ m _____ cm

780 m + 390 m = _____ km _____ m

400 ml + 750 ml = _____ ℓ _____ ml

9 in. + 6 in. = _____ ft _____ in.

8 oz + 10 oz = _____ lb _____ oz

3 ft 7 in. + 10 in. = _____ ft _____ in.

8 h 30 min + 20 min = _____ h _____ min

3 ft 7 in. + 6 ft 10 in. = _____ ft _____ in.

7 km 900 m + 7 km 105 m = _____ km _____ m

Add and Subtract Measures - Practice

1. Add or subtract in compound units.

   (a)   15 m 84 cm + 2 m 98 cm = _____ m _____ cm

   (b)   8 ℓ 900 ml − 5 ℓ 750 ml = _____ ℓ _____ ml

   (c)   85 m 50 cm − 32 m 75 cm = _____ m _____ cm

   (d)   32 yd − 7 yd 2 ft = _____ yd _____ ft

   (e)   45 ft 3 in. + 12 ft 8 in. = _____ ft _____ in.

   (f)   142 gal 1 qt + 457 gal 3 qt = _____ gal _____ qt

   (g)   5 kg 530 g − 4 kg 850 g = _____ kg _____ g

   (h)   4 lb 4 oz + 21 lb 13 oz = _____ lb _____ oz

   (i)   3 lb 7 oz − 10 oz = _____ lb _____ oz

   (j)   5 ℓ 22 ml + 2 ℓ 456 ml = _____ ℓ _____ ml

   (k)   6 qt 2 c − 4 qt 3 c = _____ qt _____ c

   (l)   12 weeks 4 days − 4 weeks 6 days = _____ weeks _____ days

   (m)   18 h 15 min − 3 h 40 min = _____ h _____ min

2. Carlie is 3 ft 7 in. tall and her little sister, Sondra, is 33 in. tall. What is the difference in their heights?

3. A tank has 10 ℓ 500 ml of water. Container A has a capacity of 4 ℓ 900 ml, Container B has a capacity of 3 ℓ 400 ml, and Container C has a capacity of 2 ℓ 750 ml. John pours water from the tank into container A until it is full, then into container B until it is full, and the rest into container C. How much more water can container C hold?

Add and Subtract Measures - Answers

1. Add or subtract in compound units.

    (a) 15 m 84 cm + 2 m 98 cm = **18** m **82** cm

    (b) 8 ℓ 900 ml − 5 ℓ 750 ml = **3** ℓ **150** ml

    (c) 85 m 50 cm − 32 m 75 cm = **52** m **75** cm

    (d) 32 yd − 7 yd 2 ft = **24** yd **1** ft

    (e) 45 ft 3 in. + 12 ft 8 in. = **57** ft **11** in.

    (f) 142 gal 1 qt + 457 gal 3 qt = **600** gal **0** qt

    (g) 5 kg 530 g − 4 kg 850 g = **0** kg **680** g

    (h) 4 lb 4 oz + 21 lb 13 oz = **26** lb **1** oz

    (i) 3 lb 7 oz − 10 oz = **2** lb **13** oz

    (j) 5 ℓ 22 ml + 2 ℓ 456 ml = **7** ℓ **478** ml

    (k) 6 qt 2 c − 4 qt 3 c = **1** qt **3** c

    (l) 12 weeks 4 days − 4 weeks 6 days = **7** weeks **5** days

    (m) 18 h 15 min − 3 h 40 min = **14** h **35** min

2. Carlie is 3 ft 7 in. tall and her little sister, Sondra, is 33 in. tall. What is the difference in their heights?

    Sondra: 33 in = 2 ft 9 in.
    3 ft 7 in. − 2 ft 9 in. = **10 in.**
    The difference in height is 10 in.

3. A tank has 10 ℓ 500 ml of water. Container A has a capacity of 4 ℓ 900 ml, Container B has a capacity of 3 ℓ 400 ml, and Container C has a capacity of 2 ℓ 750 ml. John pours water from the tank into container A until it is full, then into container B until it is full, and the rest into container C. How much more water can container C hold?

    Total capacity of the 3 containers: 4 ℓ 900 ml + 3 ℓ 400 ml + 2 ℓ 750 ml = 11 ℓ 50 ml
    11 ℓ 50 ml − 10 ℓ 500 ml = **550 ml**
    Container C can hold 550 ml more water.

Multiply Measures - Practice

1. Fill in the blanks.

   (a) 4 kg 201 g x 5 = _____ kg _____ g

   (b) 4 ℓ 230 ml x 9 = _____ ℓ _____ ml

   (c) 12 m 62 cm x 8 = _____ m _____ cm

   (d) 8 min 15 s x 5 = _____ min _____ s

   (e) 6 h 20 min x 6 = _____ h _____ min

   (f) 2 ft 10 in. x 5 = _____ ft _____ in.

   (g) 3 lb 8 oz x 4 = _____ lb _____ oz

   (h) 16 gal 3 qt x 6 = _____ gal _____ qt

   (i) 3 ft 9 in. x 12 = _____ ft _____ in.

   (j) 4 gal 3 c x 7 = _____ gal _____ c

2. The side of a square picture frame measures 1 ft 4 in. What is its perimeter?

3. Maryellen bought 10 cans of soup. Each can weighed 1 lb 4 oz. What was the total weight?

4. One week Josh jogged for 1 h 25 min every day on Monday through Saturday, and for 40 minutes on Sunday. How much time did he spend jogging that week?

5. A seamstress bought some cloth. She cut 4 pieces of cloth to make some shirts. Each piece was 2 m 400 cm long. There was 1 m 800 cm left over. What was the length of the cloth she bought?

Multiply Measures - Answers

1. Fill in the blanks.

   (a) 4 kg 201 g x 5 = **21** kg **5** g

   (b) 4 ℓ 230 ml x 9 = **38** ℓ **70** ml

   (c) 12 m 62 cm x 8 = **100** m **96** cm

   (d) 8 min 15 s x 5 = **41** min **15** s

   (e) 6 h 20 min x 6 = **38** h **0** min

   (f) 2 ft 10 in. x 5 = **14** ft **2** in.

   (g) 3 lb 8 oz x 4 = **14** lb **0** oz

   (h) 16 gal 3 qt x 6 = **100** gal **2** qt

   (i) 3 ft 9 in. x 12 = **45** ft **0** in.

   (j) 4 gal 3 c x 7 = **29** gal **5** c

2. The side of a square picture frame measures 1 ft 4 in. What is its perimeter?

   1 ft 4 in x 4 = 4 ft 16 in. = 5 ft 4 in.
   The perimeter is 5 ft 4 in.

3. Maryellen bought 10 cans of soup. Each can weighed 1 lb 4 oz. What was the total weight?

   1 lb 4 oz x 10 = 10 lb 40 oz = 12 lb 8 oz
   The total weight was 12 lb 8 oz.

4. One week Josh jogged for 1 h 25 min every day on Monday through Saturday, and for 40 minutes on Sunday. How much time did he spend jogging that week?

   1 h 25 min x 6 = 8 h 30 min
   8 h 30 min + 40 min = 9 h 10 min
   He spent 9 h 10 min jogging that week.

5. A seamstress bought some cloth. She cut 4 pieces of cloth to make some shirts. Each piece was 2 m 400 cm long. There was 1 m 800 cm left over. What was the length of the cloth she bought?

   2 m 400 cm x 4 = 9 m 600 cm
   9 m 600 cm + 1 m 800 cm = 11 m 400 cm
   She bought 11 m 400 cm of cloth.

Shapes for Lesson 4.1b

Centimeter Graph Paper

Square Dot Paper

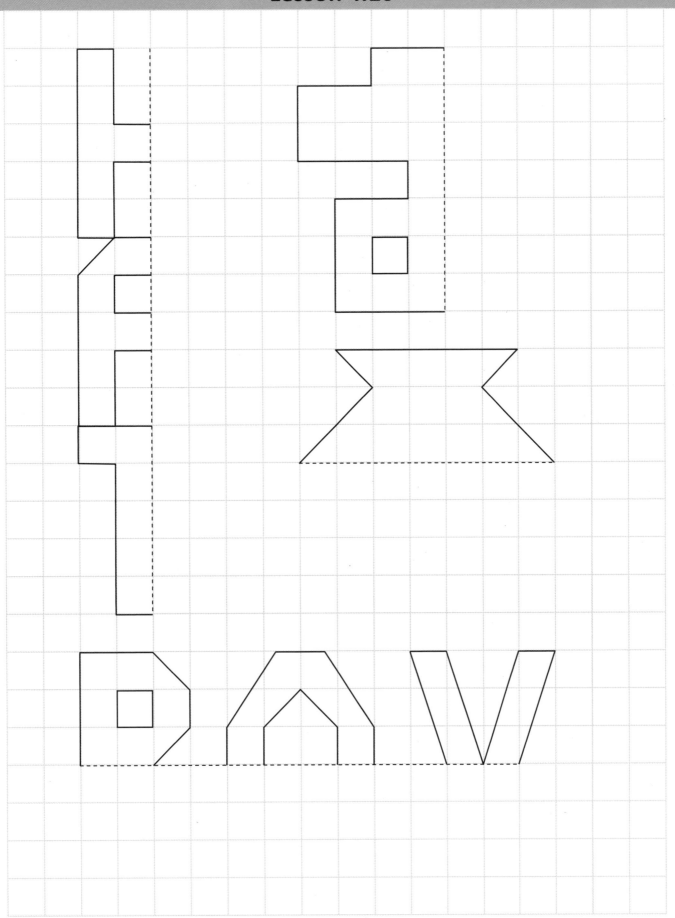

Lesson 4.1c - Enrichment

Isometric Dot Paper

Lesson 5.1a

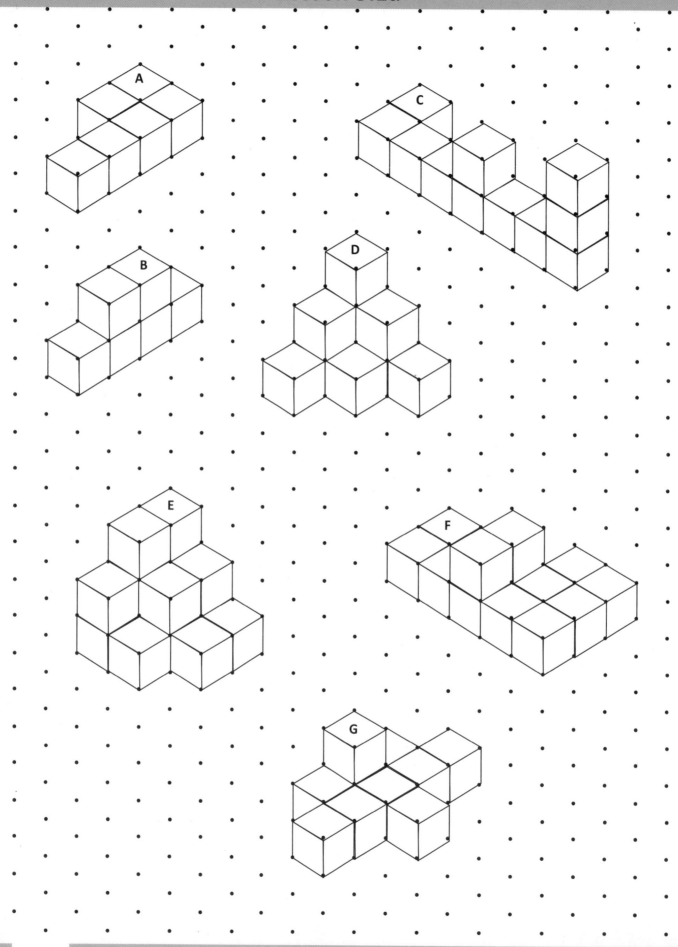

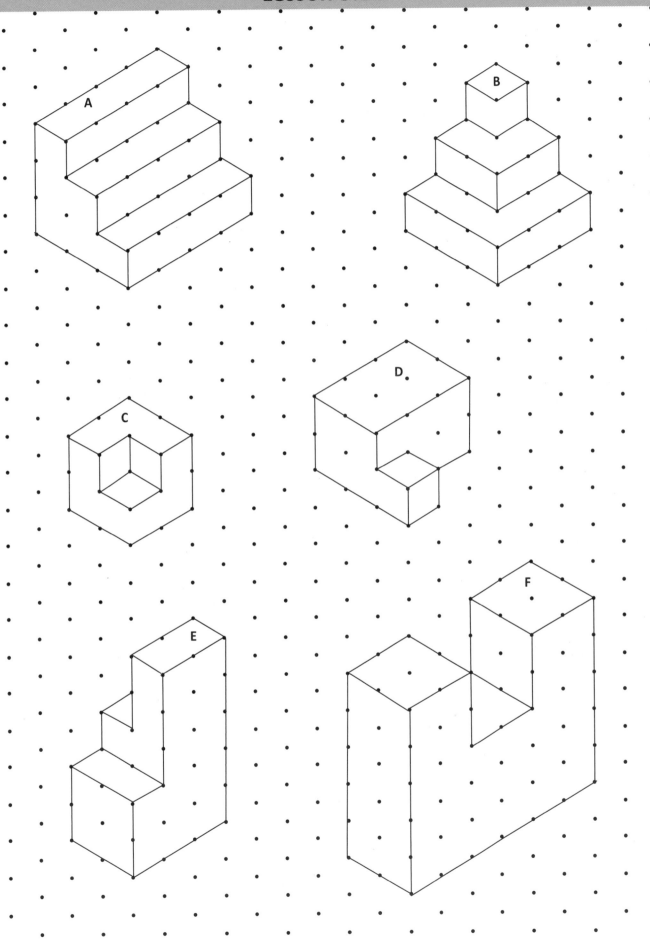

Lesson 5.1c

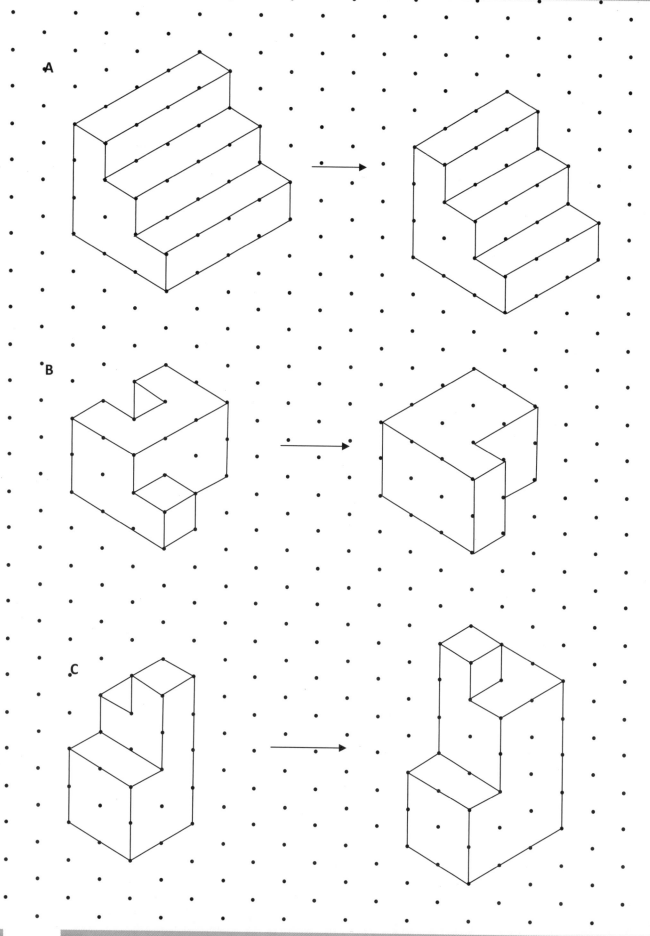

A

B

C

Net For a Cubic Inch

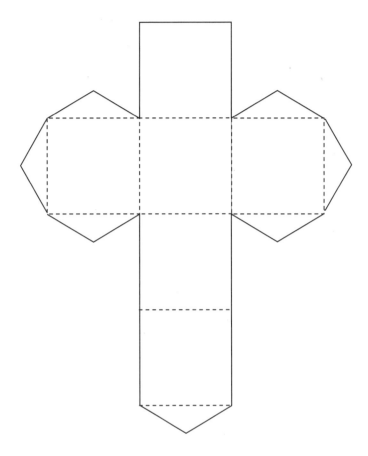

Cut along solid lines, fold along dotted lines, and use triangular tabs to glue or tape to form a cube.

1. Find the volume of a box with a length of 15 cm, a width of 12 cm, and a height of 4 cm.

2. Find the volume of a box that is 11 cm by 2 cm by 8 cm.

3. Find the volume of a 5-centimeter cube.

4. A rectangular prism is made from 2-centimeter cubes. Its dimensions are 10 cubes by 8 cubes by 4 cubes. What is its volume?

5. A rectangular container is 11 cm long, 11 cm wide, and 9 cm high. How many 2-centimeter cubes can it hold?